Cybercrime, Cyberterrorism, and Cyberwarfare

Cybercrime, Cyberterrorism, and Cyberwarfare

Crime, Terror, and War without Conventional Weapons

Robert T. Uda, MBA, MS

Contributing Authors:
Rhonda Chicone, PhD (ABD)
Bill Shervey, MS, MBA
Darrin Todd, MS

To order additional copies of this book, contact:
Xlibris Corporation
1-888-795-4274
www.Xlibris.com
Orders@Xlibris.com
63455

Contents

Dedication

Cybercrime, Cyberterrorism, and Cyberwarfare is dedicated to Dr. Robert Statica, co-founder, president, and CEO of SIG Homeland Security, LLC, for his pioneering efforts in combating cybercrime, cyberterrorism, and cyberwarfare.

Preface

*C*ybercrime, Cyberterrorism, and Cyberwarfare provides a basic understanding of cybercrime, cyberterrorism, and cyberwarfare and how to combat criminals, terrorists, and state warriors who do not use conventional weapons of bombs, grenades, and bullets. Instead, they use electrons, computers, malicious software or code, networks, and the Internet, World Wide Web, or cyberspace to transport their malware of viruses, bots, worms, logic bombs, spyware, and Trojan horses to destroy or render ineffective our computers, networks, and critical infrastructure and key resources.

Cybercrime, Cyberterrorism, and Cyberwarfare discusses ways of thwarting threat agents, intruders, hackers, hactivists, phreaks, and zombies to combat denial of service, distributed denial of service, sniffers, botnets, spoofing, e-mail spoofing, vishing, pharming, phishing, trapdoors, and spamming. *Cybercrime, Cyberterrorism, and Cyberwarfare: Crime, Terror, and War without Conventional Weapons* also covers ways of guarding against cyber threats and cyber attacks using encryption, firewalls, firmware, intrusion detection, vulnerability management, risk management, and information systems security.

If you disagree with anything that I have written in this book, I encourage you to write me and voice your disagreement. I always like to hear and learn about other people's views on whatever I write. Never do I believe that I know all truth on anything. I am always willing to change my views if someone comes up with contrary responses that make sense to me. That being said, I look forward to hearing from you.

All writings and opinions in this book are solely mine. Any error would be my error only. If you find errors, please bring them to my attention. We will correct them in subsequent editions of this book. I hope you enjoy the real-life stories in this book as I thoroughly have enjoyed living and writing about them. Thank you.

Robert T. Uda
San Marcos, California
September 2009

Chapter 1
Cybercrime, Cyberterrorism, and Cyberwarfare in Perspective

Robert T. Uda, MBA, MS

"Nearly every day our nation is discovering new threats and attacks against our country's networks. Inadequate cybersecurity and loss of information has inflicted unacceptable damage to U.S. national and economic security. The president of [the] United States must know what these threats are and how to respond to them."[1]

CSIS Commission on Cybersecurity
for the 44th Presidency

In this introductory chapter, I present a broad coverage of cybercrime, cyberterrorism, and cyberwarfare. Distinct dividing lines do not separate these three areas. Cybercrime overlaps some of cyberterrorism, and cyberterrorism overlaps some of cyberwarfare. Hence, most chapters may talk about each of these cyber security areas in an intermingled fashion.

The Center for Strategic and International Studies (CSIS) Commission report indicated three major findings as follows: "(1) cybersecurity is now a major national security problem for the United States, (2) decisions and actions must respect privacy and civil liberties,

---

[1] CSIS Commission (2008, December). *Securing cybersecurity for the 44th presidency: A report of the CSIS commission on cybersecurity for the 44th presidency*. Washington, DC: Center for Strategic and International Studies (CSIS), 96 pp.

and (3) only a comprehensive national security strategy that embraces both the domestic and international aspects of cybersecurity will make us more secure."[2]

The Obama administration has made counteracting threats to the United States' computer networks a top priority. President Obama recently called such attacks "one of the most serious economic and national security challenges" facing the United States. Cyber attacks on federal government agencies have been increasing, and thousands of federal networks have been breached including those of the Department of Homeland Security (DHS) reported security officials (Gorman, 2009).[3]

The first-of-its-kind system to protect the U.S. government's computer networks from cyberspies is being delayed by technical limitations and privacy concerns. The latest complete version of the system, known as Einstein, won't be installed for 18 months. Einstein does not protect networks from attack. It just raises the alarm after an attack has occurred. The cost is expected to exceed $2 billion. DHS spokeswoman Amy Kudwa described the various rollouts as "incremental improvements" designed to protect privacy and civil liberties. She said, "We don't want to let the perfect be the enemy of the good."[4] Indeed, perfection is the enemy of good enough.

Cybersecurity is one of the greatest security challenges the United States faces in a new and more competitive international environment in the 21st century. We face a long-term, strategic challenge in cyberspace from foreign intelligence agencies/militaries, criminals, and terrorists. Losing this struggle will wreak havoc on the economic health and national security of the United States. A general officer said in his briefing, "In cyberspace, the war has begun."[5]

---

[2] Ibid.

[3] Gorman, S. (2009, July 3). Troubles plague cyberspy defense. *The Wall Street Journal*. Retrieved July 3, 2009, from *http://online.wsj.com/article/SB124657680388089139.html*.

[4] Ibid.

[5] CSIS Commission (2008, December). *Securing cybersecurity for the 44th presidency: A report of the CSIS commission on cybersecurity for the 44th presidency*. Washington, DC: Center for Strategic and International Studies (CSIS), 96 pp.

Cybercrime

What is Cybercrime? Online activities are as vulnerable to common everyday crimes and can just as effectively compromise personal safety. Legislators, police, and citizens need to know how to protect themselves as well as those for which they are responsible. The cybercrimes listed below represent crimes that have existed long before the advent of the computer and Internet. The only difference is the tools used to commit those crimes:[6]

- Assault by threat through emails, videos, or phones
- Child pornography over the Internet using computers
- Cyber contraband where illegal items are transferred through the Internet
- Cyberlaundering where illegal funds are transferred over the Internet
- Cyberstalking using emails, phones, text messaging, webcams, websites, and/or videos to stalk and harass someone
- Cyberterrorism using computer technology to commit violence against civilians
- Cybertheft through using the computer to steal through activities related to breaking and entering, cache poisoning, embezzlement and unlawful appropriation, espionage, identity theft, fraud, malicious hacking, plagiarism, and piracy.
- Cybertresspassing, which is hacking for the purpose of entering an electronic network without permission
- Cybervandalism where computer technology is used to damage and/or destroy data

Are Cybercrime Laws Uniform? Victoria Roddel said, "You may think the penalties for breaking into a home or car (personal property) and breaking into a home computer or laptop (personal property) would be the same at the same geographic location. Not necessarily." Cybercrime either requires a computer to complete the crime or a

---

[6] Roddel, V. (2009, April 13). Cybercrime defined. Part 1 of 3 in the series: What is cybercrime? *BrightHub.com*. Retrieved July 11, 2009, from *http://www.brighthub.com/internet/security-privacy/articles/3435.aspx*.

computer is the target of a crime. However, not all countries recognize cybercrimes as crimes. Therefore, all governments should review current laws to assure that they are updated to include cybercrime to protect citizens.[7]

Cybercrime Laws Vary. Many countries still do not effectively deter, stop, limit, or punish cybercrime. The partial list below shows examples of a few countries with laws related to particular cybercrimes:[8]

- Australia—theft related to a computer system
- Belgium—sabotage
- Canada—forgery, intrusion, and theft
- Chile—child pornography
- Czech Republic—hacking may not be a cybercrime but illegal use of the accessed information is
- Ireland—theft and fraud
- Japan—forgery and intrusion
- Peru—forgery
- Spain—fraud and theft
- United Arab Emirates—using electronics to insult any religion
- USA—robot-networks (or botnets) and identity theft are now federal offenses

PCs Next in Cyber Attacks. North Korea was a prime suspect for launching recent cyber attacks. However, that isolated rogue state was not on a list of websites from the five countries the South Korean Communications Commission (KCC) said the cyber attacks may have originated. The KCC stated that the host websites believed to be behind the original attacks were based in Germany, Georgia, Austria, South Korea, and the United States. The location of the hackers behind the attacks

---

[7]   Roddel, V. (2009, May 4). Are cybercrime laws uniform? Part 2 of 3 in the series: What is cybercrime? *BrightHub.com*. Retrieved July 11, 2009, from *http://www.brighthub.com/internet/security-privacy/articles/8821. aspx.*

[8]   Roddel, V. (2009, May 4). Cybercrime laws vary. Part 3 of 3 in the series: What is cybercrime? *BrightHub.com*. Retrieved July 11, 2009, from *http:// www.brighthub.com/internet/security-privacy/articles/8822.aspx.*

is still unknown. Some analysts question North Korea's involvement, saying that it may be the work of industrial spies or pranksters.[9]

The cyber attacks saturated target websites with access requests generated by malicious software planted on PCs. These requests overwhelmed some target websites and slowed server response to legitimate traffic. Experts said that the distributed denial of service (DDoS) hacking attack spreads viruses on PCs, which turned them into zombies that, unbeknownst to the owners, simultaneously connect to specific websites. U.S. officials would not speculate on who may be responsible for the attacks. However, they noted that U.S. government websites face attacks or scams "millions of times" per day.[10]

Ahnlab, a leading South Korean web security firm, has closely examined the recent attacks and said that the new phase of attacks would target data on tens of thousands of infected PCs. Lee Byung-cheol of Ahnlab said, "The affected computers will not be able to boot and their storage files will be disabled."[11] About a year ago, this is exactly what happened to my PC. The computer locked up, and I couldn't get the computer to reboot. My hard drive was totally trashed. The computer manufacturer had to replace my hard drive, and I lost about two months worth of unsaved data.

Phishing Meet Vishing. The terms *voice* and *phishing* merge to form the term *vishing*. Phishing uses social engineering to obtain personal identifying information such as credit card numbers, bank-account numbers, personal identification numbers (PINs), social security numbers, and such kinds of information. Normally, criminals perform phishing through the Web and email. Criminals perform vishing over the phone. Caller identification (ID) spoofing is at the core of vishing. Criminals, crackpots, scammers, and bored teenagers easily feed misleading information into the caller ID system. According to the Federal Bureau of Investigation (FBI) vishing attacks are on the rise. Instead of

---

[9]   Anonymous (2009, July 10). Newsfront: PCs next in cyberattacks, says South Korea. *Newsmax.com*. Retrieved July 10, 2009, from *http://www. newsmax.com//newsfront/new_cyber_attacks/2009/07/10/234062.html.*

[10]   Ibid.

[11]   Ibid.

just stealing money, vishers are making mischief . . . potentially deadly mischief.[12]

Potentially, vishing presents a huge cyberterrorism threat. Fake calls and social engineering that send our emergency responders to the incorrect location provide terrifying implications. As with most cyber-security and cyberterrorism issues, awareness, vigilance, and a tab of paranoia may well provide the best defense.[13]

Cyberwar is a wildly asymmetrical battle. Our enemies use weapons that are the same products bought by teenagers in Wal-Mart. The enemy's cost to arm is inexpensive, yet our cost to defend is beyond our imagination. Every iPod, phone, and camera represents a potential threat to our national security. Instead of paranoia, that's just a fact of life today. That's something to keep you awake at night.[14]

European Cyber-gangs Target Small U.S. Firms. Organized cyber-gangs in Eastern Europe increasingly prey on small and mid-size companies in the United States. They have set off a multimillion-dollar online crime wave that worries the nation's largest financial institutions. A task force that represents the financial industry sent out an alert in August 2009, which outlines the problem and urges its members to implement numerous precautions now used to detect consumer bank and credit card fraud. The confidential alert says, "In the past six months, financial institutions, security companies, the media and law enforcement agencies are all reporting a significant increase in funds transfer fraud involving the exploitation of valid banking credentials belonging to small and medium sized businesses."[15]

---

[12]   Gewirtz, D. (2009, Spring). Digital defense: When your phone turns against you. *Counter Terrorism: The Journal of Counterterrorism and Homeland Security International, 15*(1), pp. 60-65.

[13]   Ibid.

[14]   Ibid.

[15]   Krebs, B. (2009, August 25). European cyber-gangs target small U.S. firms, group says. *The Washington Post*. Retrieved August 28, 2009, from *http://www.washingtonpost.com/wp-dyn/content/article/2009/08/24/ AR2009082402272_pf.html*.

Because the targets tend to be smaller, the attacks have attracted little of the notoriety that follows larger-scale breaches at big retailers and government agencies. Yet, the industry group said that some companies have suffered hundreds of thousands of dollars or more in losses. Many firms have come forward to tell their tales. In July 2009, a school district near Pittsburgh sued to recover $700,000 that cyber-gangs took from it. In May, a Texas company was robbed of $1.2 million. An electronics testing firm in Baton Rouge, Louisiana, was bilked for nearly $100,000. The FBI is working to stem the problem. Steven Chabinsky, deputy assistant director for the Bureau's Cyber Division, said, "We share a mutual concern with respect to criminals' unrelenting intent to target our nation's financial sector and customers, whether through computer hacking or by other schemes to steal customer account information and make unauthorized withdrawals."[16]

Cyberterrorism

In the summer of 2009, Paul Davis, contributing editor of the International Association for Counterterrorism & Security Professionals (IACSP) interviewed Richard L. Cañas, the New Jersey Homeland Security Director. In the interview, Mr. Cañas discussed the four mega-disasters that concern him. These mega-disasters include hurricanes, a nuclear attack, a pandemic, and cyberterrorism . . . in that order. Hence, cyberterrorism is a major concern of homeland security professionals in the United States.[17]

What is Cyberterrorism? According to the FBI, *cyberterrorism* is any "premeditated, politically motivated attack against information, computer systems, computer programs, and data which results in violence against non-combatant targets by sub-national groups or clandestine agents." Unlike a nuisance virus or a computer attack that results in a denial of service (DoS), a cyberterrorist attack causes physical violence or great financial harm. According to the U.S. Commission of Critical Infrastructure Protection, examples of cyberterrorist targets include the

---

[16] Ibid.
[17] Davis, P. (Fall 2009). An IACSP Q&A with Richard L. Cañas, the New Jersey homeland security director. *Counter Terrorism: Journal of Counterterrorism & Homeland Security International, 15*(3), 16-18.

banking industry, military installations, power plants, air traffic control (ATC) centers, and water systems. Cyberterrorism is sometimes referred to as electronic terrorism or information war.[18]

Increasing Cyberterrorist Threat. As businesses move from proprietary networks to web-based systems, they become increasing targets for terrorist attacks, according to security chiefs at the British Petroleum Company (BP). Rob Martin, manager of digital security services said, "Terrorism will increase. There's been a lot of hype about cyberterrorism, and, in a sense, it's been dismissed as a threat—but we have to look at how society has changed. Young terrorists have grown up with computers, and we've seen society become reliant on technology. They will use this against us."[19]

British Petroleum said that it was difficult to monitor underground groups. However, Martin said that intelligence services informed the company how often cyberterrorism is being mentioned so that it can accordingly adjust its threat alert. Martin said that BP was more likely to be affected by cybercrime than cyberterrorism. However, the impact of terrorism was "likely to be far more significant" than cybercrime if it did occur.[20]

Critical Infrastructure at Risk of Cyber Attack. We need to know three things about cyber attack threats:

1. The threat is real
2. There will be more than one attack
3. Attacks will be directed at physical, critical infrastructure

It will hurt if even just one attack gets through. Think for a moment about what constitutes critical infrastructure. Broadly speaking, critical infrastructure elements include the following: (1) electrical

---

[18] Anonymous (2007, June 5). Cyberterrorism. *SearchSecurity.com.* Retrieved July 15, 2009, from *http://searchsecurity.techtarget.com/ sDefinition/0,sid14_gci771061,00.html.*

[19] Espiner, T. (2007, June 22). BP: Cyberterrorism threat is increasing. *ZDNet UK.* Retrieved July 15, 2009, from *http://www.industrialdefender.com/ general_downloads/news_industry/2007.06.22_cyberterrorism_threat_ increasing.pdf.*

[20] Ibid.

power generation and distribution, (2) petrochemical production and distribution, (3) telecommunications, (4) the water supply, (5) food and agriculture, (6) hospitals and other healthcare services, (7) the transportation network, (8) law enforcement, and (9) the financial system.[21]

Because of its long reach, cyberterrorism becomes more and more scary. In the days before the Internet, a well-trained terrorist cell could disrupt power distribution by gaining physical access to the power station, plant a bomb, and take out that power station. However, with Internet access, a single terrorist hacker from across the world could eliminate the entire U.S. power grid. The Internet becomes a force multiplier for terrorist organizations. An Internet attacker incurs no personal, physical risk and could attack multiple facilities.[22]

As demonstrated on 9/11, even the Pentagon is not safe from physical attack. Furthermore, in November 2008, the Pentagon reported to Fox News that an adversary had cyber attacked its computer system in the form of virus or worms that spread rapidly through a number of military networks. Immediately after the attack, the Pentagon banned using external hardware devices, e.g., flash drives. Therefore, be careful, be smart, and pay attention to inherent, online risks.[23]

Secret European Project to Battle Online Jihad. The United Kingdom (UK) collaborates with the German, Dutch and Czech governments on a secret research project to block distribution of Islamic extremist online material. Officials throughout Europe express concerns about cyberterrorists hosting most *jihadi* websites outside of the European Union (EU). Hence, such websites cannot be eliminated from the Internet. These governments will explore technical measures such as filtering technologies plus cooperate internationally on removal notices issued to Internet Service Providers (ISPs). The European Commission will fund a project called "Exploring the Islamist Extremist Web of

---

[21] Gewirtz, D. (2009, Summer). How critical infrastructure is at risk of a cyber attack. *Counter Terrorism: The Journal of Counterterrorism and Homeland Security International, 15*(2), p. 5.

[22] Ibid.

[23] Ibid.

Europe—Analysis and Preventive Approaches," which is led by the German interior ministry.[24], [25]

Using Twitter and Facebook to Prevent Attacks Before They Happen. If counterterrorism, homeland security, and law enforcement actively conduct a close watch on social networks such as Twitter, Facebook, and MySpace, they might prevent an attack, a crime, or a mass murder. That said, however, scanning Twitter feeds is not a license to be intrusive or to harass citizens unnecessarily. We should not be arbiters of political correctness or even to police obnoxious or even hateful postings. If we limit investigations to postings that are truly representative of potential trouble, law enforcement will more likely be able to pursue this technique to prevent man-induced disasters. However, overuse it, overly harass citizens, or get on ordinary American's case for every moronic statement made, then, at some point, access to these tools for investigations would probably be curtailed by the courts. Therefore, we should use Twitter, Facebook, MySpace, and other social networking tools mainly to keep America safe, but we must also play fair.[26]

Cyberwarfare

U.S. Military Command for Cyberwarfare. In June 2009, the U.S. Department of Defense announced the establishment of a new "cyber command." The U.S. Cyber Command is designed to wage digital or cyberwarfare and to strengthen defenses against growing threats to its computer networks. Secretary of Defense Robert Gates formally established the command—the country's first—that would operate

---

[24]   Anonymous (2009, Summer). Secret European project to battle online jihad. *Counter Terrorism: The Journal of Counterterrorism and Homeland Security International, 15*(2), p. 5.

[25]   Williams, C. (2009, April 8). Secret European project to battle online jihad: 'Significant' international bid to block extremism. *The Register.* Retrieved on July 5, 2009, from *http://www.theregister.co.uk/2009/04/08/ eu_extrmism_research/print.html.*

[26]   Gewirtz, D. (Fall 2009). Using online social networks in counterterrorism and law enforcement. *Counter Terrorism: The Journal of Counterterrorism and Homeland Security International, 15*(3), pp. 8-10.

under the U.S. Strategic Command, said Pentagon spokesman Bryan Whitman. The Cyber Command begins operating in October 2009 and will be fully operational in October 2010, said Whitman.[27]

The move indicates a shift in military strategy with "cyber dominance" now part of U.S. war doctrine. Cyber dominance comes amid growing concern over the perceived threat posed by digital espionage emanating from China, Russia, and elsewhere. China has developed a sophisticated cyber warfare program. A host of intrusions in the United States and elsewhere were traced back to Chinese sources. The command will probably be located at Fort Meade, Maryland. The Pentagon will not be taking over security efforts for civilian networks from other government agencies. Lt. Gen. Keith Alexander, current director of the National Security Agency (NSA) is expected to lead the new command.[28]

Cyber-Attack. Officials say that the USAF will not only defend its computer networks, but it may also attack U.S. enemy systems. Air Force General William Lord, interim commander of the AF Cybercommand said, "Imagine what can happen to us," of an attack on Department of Defense (DoD) computers. "We'd like to take that capability and bring it down on the heads of our enemies."[29]

Admiral William Owen (USN, Ret.), former vice chairman of the Joint Chiefs of Staff said that there is little evidence that a government-wide understanding exists of the repercussions of launching a cyber attack on enemy computers. That also goes for the military. Owen co-authored the National Research Council (NRC) report titled "Technology, Policy, Law and Ethics Regarding U.S. Acquisition and Use of Cyber-Attack Capabilities." The NRC said that the report is the first comprehensive review of cyber-attack that addresses policy, legal, technical, and ethical issues.[30]

---

[27]   De Luce, D. (2009, June 23). U.S. creates military command for cyber battlefield. *AFP*. Retrieved July 2, 2009, from *http://www.google.com/hostednews/afp/article/ALeqM5jCjrZHbTK6tr91wmAX8IW24TWp_w.*

[28]   Ibid.

[29]   Magnuson, S. (2009, July). Cyber-attack: U.S. plans to destroy enemy computer networks questioned. *National Defense, XCIII* (668), pp. 22-23.

[30]   Ibid.

The report defines cyber-attack as "deliberate actions to alter, disrupt, deceive, degrade, or destroy computer systems." Cyber-attacks range from small-scale skirmishes to all-out war.[31] Gen. Lord defines potential opponents as nation states, cyber-terrorists, and common criminals. He stated that identifying the origination of an attack is very difficult.[32]

North Korea was Behind the Week of Cyber Attacks. North Korea was indeed behind the cyber attacks that targeted scores of websites in South Korea and the United States during the first week in July 2009. An anonymous Pentagon official said that the attack did not penetrate the DoD's computer systems. These systems are constantly being probed from the outside. Defense officials privately complained that the DHS was the lead agency on protecting all government agencies from cyber attacks. However, the Pentagon wasn't informed about the attacks until almost a week later after hearing about it from the media.[33]

In a DDoS attack, a large quantity of "zombie computers" conglomerated in a "botnet," or robot network, are directed such that all of them go to U.S. government websites at precisely the same time. This action knocks down less-robust websites because they cannot handle all the traffic simultaneously. Basically, it just overloads the system. Attacks on federal computer networks are common. They range from nuisance hacking to more serious assaults. Sometimes, they are blamed on China. U.S. security officials are also concerned about cyber attacks from al-Qaeda or other terrorist groups.[34]

Russian Hackers Stole U.S. IDs for Attacks. Russian hackers hijacked American identities (IDs) and U.S. software tools and used them in an attack on Georgian government websites during the war between Russia and Georgia last August 2008, according to new research released on August 17, 2009, by a nonprofit U.S. group. In addition to refashioning common Microsoft Corporation software into a cyber-weapon, hackers collaborated on popular U.S.-based social-networking sites (Twitter and Facebook) to coordinate attacks on Georgian sites, the U.S. Cyber

---

[31] Ibid.

[32] Ibid.

[33] Anonymous (2009, July 9). Pentagon official: North Korea behind week of cyber attacks. *FOXNews.com.* Retrieved July 10, 2009, from *http:// www.foxnews.com/printer_friendly_story/0,3566,530781,00.html.*

[34] Ibid.

Consequences Unit found. While the cyber attacks on Georgia were examined shortly after the events last August, these U.S. connections were previously unknown. The research indicates how cyberwarfare has outpaced military and international agreements, which don't consider the probability of American resources and civilian technology being used as weapons.[35]

Common means of attack include ID theft, social networking, and modifying commercial software; however, integrating them elevates the attack method to a new level, said Amit Yoran, a former cybersecurity chief at the DHS and now chief executive officer (CEO) of computer-security company NetWitness Corporation. In August 2008, the cyber attacks significantly disrupted Georgia's communications capabilities by disabling 20 websites for more than a week. Among the sites eliminated last year included those of the Georgian president, defense minister, National Bank of Georgia, and major news outlets. The five-day Russian-Georgian conflict in August 2008 killed hundreds of people, crushed Georgia's army, and left two segments of its territory bordering Russia—Abkhazia and South Ossetia—under Russian occupation.[36]

More Devastating than a Nuclear Bomb Explosion. Senator Kit Bond, vice chairman of the Senate Select Committee on Intelligence, told *Newsmax* that a cyber-attack on the United States could be economically more devastating than a nuclear explosion and could also cause massive deaths. Bond said, "You can cause more economic harm with a cyber-attack than a nuclear one." Furthermore, he said, "It could crush our country and the world economy, which depends upon the United States as the world's leading economy. If they take us down, they cripple everybody." The Missouri Republican said that the recent attacks demonstrate how such a cyber-attack "could take down our entire infrastructure, which depends upon the use of computers and information technology."[37]

---

[35] FoxNews.com (2009, August 17). Russian hackers stole U.S. IDs for attacks. *The Wall Street Journal*. Retrieved August 17, 2009, from *http://www.foxnews.com/printer_friendly_story/0,3566,539900,00.html*.

[36] Ibid.

[37] Kessler, R. (2009, July 15). Sen. Bond: Cyberattack more devastating than nuclear bomb. Newsmax.com. Retrieved July 15, 2009, from *http://www.newsmax.com/kessler/kit_bond_cyberattacks/2009/07/15/235753.html*.

Senator Bond opposes President Barack Obama's plan to appoint a cyber czar to oversee the U.S. response. Bond said, "I'm not willing to live under a Russian-style czarist system." "He's talking about a czar for everything. We've got a system that works. Now I'm not in favor necessarily of forming a new bureaucracy. But that responsibility ought to be in the hands of either a new agency or an existing agency, where the Senate has confirmation authority and the head of it reports to Congress to tell us what they're doing."[38]

Whether located in the DoD or DHS, this new agency would be responsible for providing assistance to the private sector. Additionally, it would work closely with the intelligence community, the military, and the rest of the government. Bond said that this agency could assure that "we have the capacity to pre-empt strikes where we see a potential threat arising, before it destroys our system."[39]

Cyberspy Defense. The flagship system designed to protect the U.S. government's computer networks from cyberspies is being stymied by technical limitations and privacy concerns, according to current and former national-security officials. The latest complete version of the system, known as Einstein, won't be fully installed for 18 months, according to officials, seven years after first rolled out. This system doesn't protect networks from attack, but it raises the alarm after an attack is in process.[40]

A more capable version has tripped privacy alarms, which could delay its rollout. Since the NSA acknowledged eavesdropping on phone and Internet traffic without warrants in 2005, security programs have been hounded by privacy concerns. For Einstein, AT&T Corp., which would test the system, has sought written approval from the Department of Justice (DOJ) before it would agree to participate. Designed to protect all nonmilitary government computers, the total cost of the system is classified, but officials familiar with the program said the price tag was expected to exceed $2 billion.[41]

---

[38]   Ibid.

[39]   Ibid.

[40]   Gorman, S. (2009, July 3). Troubles plague cyberspy defense. *The Wall Street Journal*. Retrieved July 3, 2009, from *http://online.wsj.com/article/ SB124657680388089139.html*.

[41]   Ibid.

The Obama administration has made combating threats to the nation's computer networks as a top priority. President Barack Obama recently called such attacks as "one of the most serious economic and national security challenges" facing the United States. Attacks on the government have been intensifying. Thousands of federal networks are being breached including those of the DHS.[42]

Conclusion

The remaining chapters of this book cover cybercrime, cyberterrorism, and cyberwarfare in greater detail. Cybercrime is with us on a daily basis (24 hours a day and 7 days a week). Furthermore, it will continue on as long as we use computers, networks, phones, videos, webcams, emails, and the Internet. However, the most deadly one of the three is cyberwarfare. Cyberwarfare is not being conducted against us by individual, civilian hackers or by terrorist organizations but is being conducted by nation states in an organized, methodical manner.

The United States must be prepared to conduct cyberwarfare. Indeed, we must be ready to conduct offensive cyberwarfare instead of just defensive cyberwarfare. Additionally, we must be prepared for massive cyber retaliation should our entire nation be massively attacked by China, Russia, Korea, Iran, or any other nation that would do us harm. Massive Retaliation was a national strategy that ended with the close of the Cold War and the demise of the Soviet Union. However, it should be revived for cyberwarfare, or we may see the demise of the United States. We must not allow that outcome to happen.

---

[42]   Ibid.

Chapter 2
Domestic Efforts to Combat Cyberterrorism

Robert T. Uda, MBA, MS

In this section, I evaluate whether government agencies and the private sector are for implementing counterterrorism measures to combat cyberterrorism.

Government Agencies and the Private Sector

Information Sharing. Within a year after 9/11, security expert Sharon Gaudin said, "In the cyber area, companies are the most common victims of attacks. Government can't analyze that information and determine what the trends are or if this is telling of a larger attack coming, if private companies don't share information about their attacks . . . There's still the thought that companies that have been attacked should keep the information about it internal, rather than share that information with law enforcement and government agencies . . . If they don't have a good relationship with the government, they won't get the threat information they need and their ability to respond is going to be limited."[43] Since 2002, things have improved with sharing of private sector information with the government. However, sharing among government agencies, such as among the FBI, CIA, DoD, and other agencies, still remained a problem. Establishment of the DHS in November 2002 attempted to improve information sharing.

---

[43] Gaudin, S. (2002, July 19). Security expert: US companies unprepared for cyberterror. *IT Management Security*. Retrieved November 25, 2002, from *http://itmanagement.earthweb.com/secu/print.php/1429851.*

Private Sector is Key to Solution. Although governments administer only a minority of the nation's critical infrastructure computer systems, government at all levels perform essential services in the agriculture, food, water, public health, emergency services, social welfare, information and telecommunications, defense, energy, transportation, banking and finance, chemicals, and postal shipping sectors that depend on cyberspace for their delivery. Governments can lead by example in cyberspace security including fostering a marketplace for more secure technologies through their procurement.[44] However, government cannot solve the cyber security problem for the private sector. The solution must come from the private sector.

Reforms Needed. As late as the fall of 2006, Evan Kohlmann wrote, "The United States is gradually losing the online war against terrorists. To combat terrorists, we must monitor terrorist activities online in the same way the U.S. keeps tabs of terrorists in the real world. Doing so will require a realignment of U.S. intelligence and law enforcement agencies, which lag behind terrorists organizations in adopting information technologies. At present, unfortunately, senior counterterrorism officials refuse even to pay lip service to the need for such reforms. That must change—and fast. One of the most important ways in which terrorists use the Internet is as a medium for propaganda."[45] The officials' refusal to implement reforms is a sad state of affairs.

Air Force Cyber Command. As late as September 2007, the U.S. Air Force had established a provisional Cyber Command as part of an expanding mission to prepare for wars in cyberspace. The move comes amid concerns over the vulnerability of the U.S. communications and computer networks to cyber attack in a conflict, as well as the military's desire to exploit the new medium. Air Force Secretary Michael Wynne announced the creation of the new command at Barksdale Air Force Base in Louisiana, where the Air Force's existing cyberwarfare operations are centered. Officials said the provisional command will pave the way within a year for the creation of the Air Force's first major command

---

[44] Bullock, J. A., Haddow, G. D., Coppola, D., Ergin, E., Westerman, L., & Yeletaysi, S. (2006). *Introduction to Homeland Security, Second Edition.* Oxford, United Kingdom: Elsevier Butterworth-Heinemann.

[45] Kohlmann, E. F. (2006, September/October). The real online terrorist threat. *Foreign Affairs, 85(5)*, p. 115.

devoted to cyberwarfare operations.[46] It was about time the government acted to prepare us for cyberwarfare.

Key Cyber Security Activities. The FBI estimated that computer-related crimes cost U.S. businesses $67 billion per year.[47] Consequently, the U.S. government places an especially high priority on protecting our cyber infrastructure from terrorist attack by unifying and focusing the key cyber security activities performed by the Critical Infrastructure Assurance Office (formerly part of the Department of Commerce) and the National Infrastructure Protection Center (FBI). The Directorate augments those capabilities with the response functions of the National Cyber Security Division (NCSD) and United States Computer Emergency Response Team (US-CERT).[48]

Public Access to the Internet. It is interesting to note that there is an inverse relationship between public access to the Internet and the inability of governments and institutions to control information flow and, hence, state allegiance, ideology, public opinion, and policy formulation. Increase in public access to the Internet results in an equivalent decrease in government and institutional power. Indeed, after September 11, 2001, Internet traffic statistics show that many millions of Americans have connected to alternative news sources outside the continental United States (CONUS). The information they consume can be and often is contrary to U.S. government statements and U.S. mainstream media reporting. Recognizing this, terrorists will coordinate their assaults and cleverly use cyberspace for manipulating perceptions, opinion, and the political and socioeconomic direction of many nation-states.[49]

---

[46]   Anonymous (2007, September 18). US Air Force sets up Cyber Command. *Breitbart.com*. Retrieved September 19, 20007, from *http://www.breitbart. com/print.php?id=070918201110.nyigxaco&show_article=1*.

[47]   Wagner, B. (2007, July). Electronic jihad: Experts downplay imminent threat of cyberterrorism. *National Defense*, pp 34-36.

[48]   Sweet, K. M. (2006). *Transportation and Cargo Security: Threats and Solutions.* Upper Saddle River, New Jersey: PEARSON Prentice Hall.

[49]   Stanton, J. J. (2002, February). Terror in cyberspace: Terrorists will exploit and widen the gap between governing structures and the public. *The American Behavioral Scientist, 45(6)*, 1017-1032.

Cybercrime Threatens Everyone. In conclusion, the government and private sector alike recognize that cybercrime threatens everyone. This thinking has prompted both sides to forge partnerships and increase information sharing. DHS has collaborated with private firms when it wrote a computer breach scenario in the National Response Plan. As mentioned earlier, the Air Force has reached out to the private sector as it develops its new Cyber Command, the service's future focal point for computer warfare. The AF has also signaled a desire to work with the private sector to train airmen in cyber security. Despite these collaborative efforts to protect against computer incursions, the government and private sector face significant operational problems, wrote the Government Accountability Office (GAO) in a June 2007 report. GAO criticized both sides for not always detecting or reporting Internet-based crimes and says that law enforcement organizations have difficulty retaining personnel with such expertise.[50]

U.S. Legal Requirements and Computer Forensics

In this section, I analyze the practicality of U.S. legal requirements and computer forensics in addressing cyberterrorism.

Digital Technologies. The advent of digital technology and the convergence of computing and communications change the way we live. These trends create unprecedented opportunities for crime. Criminal activities unforeseen two decades ago are facts of life today. Digital technologies provide ordinary citizens, even juveniles, with the capacity to inflict massive harm. Public prosecutors must possess the knowledge that permits an effective response. The continued advancement of digital technology creates new opportunities for criminal exploitation.[51]

Computer Forensics. Forensics use technology and science to investigate and recover facts to resolve criminal matters. Computer forensics applies technology in discovering evidence to use in criminal or civil courts of law. The process involves recovering damaged and

---

[50] Wagner, B. (2007, October). Electronic attackers: Computer crimes keep government and industry on the defensive. *National Defense*, pp 24-26.

[51] Grobsky, P. (2007, October 13). Requirements of prosecution services to deal with cybercrime. *Crime Law Soc Change, 47*, 201-223.

deleted files. In particular, some cases use computer forensics as evidence to indict criminal offenders or to locate a missing person.[52]

Cybercrimes. Traditional law enforcement ends at national borders; whereas, cybercrime does not. The perpetrators may be located anywhere in the world, and unless there is international cooperation, it is very difficult to arrest and prosecute the attackers. Unfortunately, in most countries, there is no clear basis in law for prosecuting cyber-criminals and no basis for extradition.[53]

Cyber Attacks. America's cyberspace links the United States to the rest of the world. A network-of-networks spans the planet, thereby, allowing malicious actors on one continent to act on systems thousands of miles away. Cyber attacks penetrate borders at light speed; hence, discerning the source of malicious activity is difficult. America must be capable of safeguarding and defending its critical systems and networks. Enabling the ability to do so requires a system of international cooperation to facilitate information sharing, reduce vulnerabilities, and deter malicious actors.[54]

Cyberterror. Today, most of cyberterror is plagued by very fuzzy definitional boundaries. For instance, would a cyber attack on a country's stock market be a form of cyberterror because it would involve no immediate, tangible violence? Additionally, could we reach universal agreement on the essential nature of the stock market to the survival of those living in the system? What about an attempt to corrupt information within a system such as that pertaining to blood types in a hospital? If the only result is additional costs in terms of time delay/effort and not lives lost or medical emergencies, is this still terrorism?[55] If we cannot reach consensus on answers to

---

[52]   Philpott, D. (n.d.). Special report: Computer forensics and cyber security. *Homeland Defense Journal*, 12 pp.

[53]   Cordesman, A. H. (2002). *Cyber-threats, Information Warfare, and Critical Infrastructure Protection: Defending the US Homeland*. Westport, Connecticut: Praeger Publishers.

[54]   Bullock, J. A., Haddow, G. D., Coppola, D., Ergin, E., Westerman, L., & Yeletaysi, S. (2006). *Introduction to Homeland Security, Second Edition*. Oxford, United Kingdom: Elsevier Butterworth-Heinemann.

[55]   Combs, C.C. (2006). *Terrorism in the Twenty-First Century (4th ed.)*. Upper Saddle River, New Jersey: Pearson Prentice-Hall.

these questions, how could we effectively discuss cyberterror in a meaningful way?

Digital Evidence. Whether the crime is a new one, like phishing,[56] or a more conventional fraud committed with new technology, the intangible nature of digital evidence is also a challenge for prosecutors. Assembling evidence, preserving its integrity, and presenting it in a comprehensible manner to a court has always been the prosecutor's role. However, judges and juries who are unfamiliar with high technology may be uncomfortable with evidence in digital form. Furthermore, skilled defense counsel may seek to sow the seeds of doubt by suggesting that computer malfunction or human intervention contaminated the evidence in question. The accused, moreover, may argue that it was not he or she who committed the criminal act, but rather someone else with either direct or remote access to the computer. One recognizes that capacity building is important in advanced industrial nations as well as to those on the other side of the digital divide.[57]

Prosecuting Internet Crimes. The GAO suggests that law enforcement personnel struggle to prosecute Internet crimes because there is a limited pool of highly trained specialists. Additionally, they report that law enforcement efforts are hampered by the cross-border nature of such crime. Officials have a difficult time figuring out the laws and legal procedures of multiple jurisdictions.[58]

Increased Funding. In May 2007, Congress introduced new legislation that would increase funds for law enforcement and allow the Department of Justice to impose stricter penalties for computer

---

[56] Krone, T. (2005). *Phishing, High Tech Crime Brief No. 9.* Canberra: Australian Institute of Criminology. Retrieved July 5, 2005, from *http://www.aic.gov.au/publications/htch/htch009.html. Phishing is transmitting a form of Spam containing links to Web pages that are designed to appear to be legitimate commercial sites. They seek to fool users into submitting personal, financial, or password data. Clicking on the link may also lead to infection of one's computer by a virus or may allow access to one's computer by a hacker.*

[57] Grobsky, P. (2007, October 13). Requirements of prosecution services to deal with cybercrime. *Crime Law Soc Change, 47,* 201-223.

[58] Wagner, B. (2007, October). Electronic attackers: Computer crimes keep government and industry on the defensive. *National Defense,* pp 24-26.

criminals. The House Judiciary Subcommittee on Courts, the Internet, and Intellectual Property introduced the Cybercrime Enhancement Act of 2007. If passed, the law would allocate $10 million per year to federal law enforcement through 2011. The money would be given to the U.S. Secret Service, the Attorney General's office, and the FBI to combat identity theft and other cybercrimes.[59] The main question begged is this: Is $10 million per year enough to solve this problem? This amount seems to be a mere pittance.

Persisting Problems. In conclusion, enforcing the law across borders is a real problem . . . particularly since cyberterrorism can be committed worldwide from any location on earth. Pinpointing the source is a real challenge for cyber-cops. Definitional boundaries present another hurdle to overcome. How do we communicate with other nationals if we cannot agree on the terminology we use to discuss *cyberterrorism*? The intangible nature of digital evidence is also a challenge for prosecutors, judges, and juries. Another area of weakness is that we just do not have enough people trained and knowledgeable in sufficient depth to deal with the legal requirements, computer forensics, and cyberterrorism. Congress is working to provide more funds to deal with these problems; however, it just may be too little too late. What we need is a more concerted effort to deal with this cyberterror problem on a more global, systematic approach.

A Duty of Care in Cyberspace

In this section, I analyze the topic of "A Duty of Care in Cyberspace"and how it relates to cyberterrorism.

Definition. The *Law.com Dictionary* defines *"a duty of care"* as a requirement that a person act toward others and the public with the watchfulness, attention, caution, and prudence that a reasonable person in the circumstances would use. If a person's actions do not meet this standard of care, we then consider the acts as negligent, and we may claim any damages resulting in a lawsuit for negligence.[60]

---

[59]   Wagner, B. (2007, July). Electronic jihad: Experts downplay imminent threat of cyberterrorism. *National Defense*, pp 34-36.

[60]   Anonymous (n.d.). *LAW.com Dictionary*. Retrieved March 22, 2008, from *http://dictionary.law.com/*.

The 9/11 Class-action Suit. In the class-action litigation brought by families of the September 11th victims against the airlines, airport security companies, airplane manufacturers, and the owners and operators of the World Trade Center, the court examined two main elements:

- Whether the various defendants owed *a duty of care* to the people in the World Trade Center and on the planes that crashed
- Whether the terrorist act was foreseeable

Upon finding that the case should go to a jury, the court stated that we impose a duty on a company when the relationship between the company and user requires the company to protect the user from the conduct of others. The court noted that we already depend on others to protect the quality of our water and the air we breathe. This *duty of care* extends to private companies as well.[61] So, now, how does "duty of care" relate to cyberspace and cyberterrorism?

Corporate Duty of Care. One of the major legal risks arising from cyber security breaches is the possibility of derivative suits against corporate officers and directors alleging that they have breached their *duty of care* by failing adequately to protect against security breaches. Directors and officers have a fiduciary obligation to use reasonable care in overseeing the business operations of the company, under the doctrine of *corporate duty of care*. See, e.g., In re Logue Mechanical Contracting Corp., 106 B.R. 436, 439 (Bankr. W.D. Pa. 1989). Traditionally, directors and officers could defend against *a duty of care* claim by showing that they acted with reasonable care by relying on information reasonably available to them. In the past few years, however, courts have expanded this reasonable care standard to create *a duty of oversight* requiring directors and officers to act affirmatively to assure that adequate information and compliance systems are in place.[62]

---

[61]   Cook, W. (2004, May). A foreseeable future: For liability purposes, the courts have declared terrorism to be a predictable security threat. CSOs need to adapt if they want to survive. *CSO Magazine*. Retrieved March 22, 2008, from *http://www.csoonline.com/read/050104/flashpoint.html*.

[62]   Matus, W. C., Polak, V. L., Mancini, A. J. P., & Nonna, J. M. (2002, March 11). Now more than ever, cyber security audits are key. *The National Law Review*, p. C8-C10.

Private Sector Involvement and Funding. Next, we look at how the private sector must be involved in fighting cyberterrorism. Christopher Beggs and Matthew Butler wrote that organizations like the U.S. Department of Homeland Security, AusCert [in Australia], terrorism treaties, and other anti-cyberterrorism organizations need funding. This funding also needs to come from the private sector, as it too possesses *a duty of care* to people worldwide. However, they also must be a primary contributor to cyberterrorism policies, strategies, and technologies.[63]

Software Developers. Software developers must consider security quality as they develop their products. Jennifer Chandler said that the negligence analysis requires that one consider the tradeoffs inherent in pressuring software developers to improve security quality. Should all, some, or no software developers owe *a duty of care* to users of cyberspace? What is the standard of reasonable care in software development? Negligence law does not demand perfection, and society would likely be unable to afford it. However, we may find certain forms of error or software development practices to be negligent.[64]

Hard Drives Disposal. Businesses must also take better care of disposing of old computer hard drives. According to a study conducted by a security firm (Rits Information Security), businesses are leaving personal information, including credit card numbers, customer data, and client files, on hard drives that are sold into the second-hand market. Firms have sourced drives, examined by Rits, openly on the Internet and in online auctions. The survey looked at the information remaining on the disk that unveiled some alarming results. In one case, some 300 credit card numbers from an

---

[63]   Khosrowpour, M. (Ed., 2004). Beggs, C., & Butler, M. (2004). Developing New Strategies to Combat Cyberterrorism, pp. 388-390. *Innovations Through Information Technology, Volume 1.* 2004 IRMA [Information Resources Management Association] International Conference. Idea Group Inc (IGI), 1,458 pp.

[64]   Chandler, J. A. (2005, September 8-10). Safety & security in a networked world: Balancing cyber-rights & responsibilities. An Oxford Internet Institute Conference. Retrieved March 22, 2008, from *http://www.oii.ox.ac. uk/microsites/cybersafety/?view=programme&day=8&expand=yes.*

organization involved in fundraising for a large charity event were present on one disk, while customer data from a major Irish bank was on another.[65]

Disposing of Information. "Most people are not aware of the implications of pressing delete, doing a simple format, or overwriting the operating system," said Vivienne Mee, Rits Security, speaking with *ENN*. "Home users in particular aren't aware, but large organizations should be. The study did show that neither [parties] are using methods to securely dispose of information." Failures to do so could leave firms open to action under Data Protection legislation. "They are in breach of legislation," said Mee. "They have a duty of care."[66]

Online Forum Managers. In C-032986/03 Moshe Boshmitz v. Anat Aronowitz, Magistrates Court of Tel-Aviv Jaffa, Israeli Judge Shoshana Almagor held that the manager of an online forum (Ms. Anat Aronowitz) might be liable for the content published by the forum users on a theory of negligence. The Court held that it is "clear beyond doubt" that a forum manager (Ms. Aronowitz) has "*conceptual duty of care*"toward the Claimant (Dr. Moshe Boshmitz) in that she should have anticipated that harm might come to him due to the messages and is, therefore, liable for her content. The determination that the forum manager has *a duty of care* towards its forum users, to the extent of imposing liability for a failure to delete posts, establishes a greater responsibility for such function and may have a chilling effect on forum managers (particularly those who mostly perform this function voluntarily and as a hobby).[67]

Safe Working Environment. Every organization has *a duty of care* to provide a safe working environment for its employees. This duty of care not only encompasses an employee's physical safety but may also include their psychological safety and well-being. Due to the nature of a police officer's job, we can neither avoid nor remove the stressors that

---

[65] O'Brien, C. (2007, November 14). Used hard drives are ID theft paradise. *ENN* [*Electric News Net*]. Retrieved March 22, 2008, from *http://www. electricnews.net/print/10123430.html.*

[66] Ibid.

[67] Kagan, O. (2007, June 27). Internet law: Israeli court holds forum manager liable for user content. *IBLS* [Internet Business Law Services, Inc.] *Internet Law—News Portal.* Retrieved Mar 22, 2008, from *http://www.ibls.com/ internet_law_news_portal_view.aspx?s=latestnews&id=1789.*

cause emotional distress. Research, therefore, focuses on developing a holistic, proactive risk management framework that provides guidance on best practice for police agencies in fulfilling their psychological *duty of care* obligations to their employees.[68]

Conclusion. In conclusion, *a duty of care* not only applies to individual persons, but it also applies to private companies. Hence, we now have *corporate duty of care*. Besides governments funding anti-cyberterrorism, the private sector must also contribute their fair share. However, expected not only to contribute to funding, they also must contribute primarily to developing cyberterrorism policies and technologies. Software developers also must take care not to be negligent in certain forms of software errors or software development practices. Furthermore, businesses must also take great care in disposing of computer hard drives with confidential and sensitive information stored in them. Online forum managers must also take responsibility for negligent content published on their websites. They must display *a duty of care* towards their website users. Law enforcement organizations must also fulfill a *psychological duty of care* towards their police officers when they are under high emotional stress resulting from their jobs. Therefore, the concept of *a duty of care* is becoming pervasive throughout cyberspace and in the worldwide battle against cyberterrorism.

State and Local Governments

In this section, I identify what measures state and local governments are using to combat cyberterrorism.

State and Local Government Responsibilities. The 50 states, 4 territories, and 87,000 local jurisdictions that comprise the United States maintain an important and unique role in protecting our critical infrastructures (CIs) and key assets (KAs). All U.S. states and territories have established Homeland Security liaison offices to manage their counterterrorism and critical infrastructure protection (CIP) efforts. In addition, the states possess law enforcement agencies,

---

[68]   Anonymous (2002, February). Current research: Development of a risk management strategy for duty of care issues relating to high-risk operational policing. *ACPR* [Australasian Centre for Policing Research] *Bulletin, No. 11.*

National Guard units, and other critical services that protect their communities.[69]

State, county, municipal, and local governments fund and operate the emergency services that would respond in the event of a terrorist attack. Ultimately, all manmade and natural disasters are local events—with local units being the first to respond and the last to leave. Since September 11, 2001, every state and many cities and counties have addressed homeland security issues either through an existing office or through a newly created office. Many have established anti-terrorism task forces. Many have published or are preparing homeland security strategies. Some of these strategies are based on existing plans for dealing with natural disasters. Each level of government must coordinate with other levels to minimize redundancies in homeland security actions and ensure integrated efforts.[70] State and local government first responders also are responsible for dealing with the aftermath of cyberterrorism attacks.

State and Local Governments are Unprepared. In general, state and local governments have not organized effectively to deal with the changes in CI and information technology (IT). Many programs are badly underfunded and lag way behind the private sector. This shortcoming is systemic in many areas of protection capability at the prevention, mitigation, and reconstitution levels.[71] It is only a matter of time before cyberterrorists perpetrate an attack on a city or county government.[72] Only then will Congress funnel sufficient funds into the state and local counter-cyberterrorism efforts.

What State Government Could Do? Now, we address what state government could do to make our CI organizations secure and to protect

---

[69] Anonymous (2003, February). *The National Strategy for the Physical Protection of Critical Infrastructures and Key Assets.* Washington, DC: The White House.

[70] Anonymous (2002, July). *National Strategy for Homeland Security.* Washington, DC: Office of Homeland Security.

[71] Cordesman, A. H. (2002). *Cyber-threats, Information Warfare, and Critical Infrastructure Protection: Defending the US Homeland.* Westport, Connecticut: Praeger Publishers.

[72] Misra, S. (2003, June). High-tech terror: Cities and counties need plans to respond to criminal efforts to destroy government computer networks and data. *The American City & County, 118(6),* p. HS6.

our citizens. First, we clearly need to define, through effective leadership and with a clear vision, what needs to be accomplished. Second, we need to address the need for a coordinated state strategy and a comprehensive threat assessment managed under the Lieutenant Governor's office.[73] The state government must be the driving force behind these efforts. The national government will not make it happen on the state level lest it be accused of meddling. With the exception of New York City and Los Angeles, most local governments do not possess the wherewithal to make it happen. Hence, the responsibility rests in the lap of the state government.

Cyberterrorism—Top-three Concerns. Biological attacks, chemical attacks, and cyberterrorism are the top three terrorism concerns for communities in the aftermath of the September 11, 2001, terrorist attacks. A survey carried out by the National League of Cities (NLC), Washington, DC, showed that the international terrorism threat has increased municipal officials' security concerns. Many respondents indicated that they have not developed response plans, but all have developed greater cooperation with counties, civic organizations, nonprofit organizations, other municipalities, and even the media.[74]

Local Governments—a Growing Concern. Local governments constitute a growing concern among security experts. According to the NLC survey, only 43 percent of large cities and 26 percent of all cities have developed strategies to address cyberterrorism.[75] For local governments, protecting networks and computer systems is costly and time-consuming. In addition to expanding firewalls and intrusion detection, IT departments are busy installing anti-malware software to protect against pervasive worms and viruses as well as automatic

---

[73] Kobus, W. S., Jr. (2001, November). Executive guide for state governors: State plan for securing critical infrastructure private companies. Raleigh, North Carolina: Total Enterprise Security Solutions, LLC, 19 pp.

[74] Anonymous (2002, November). International threats hit home for local leaders. *The American City & County, 117(16)*, pp. S12 & S14.

[75] Misra, S. (2003, June). High-tech terror: Cities and counties need plans to respond to criminal efforts to destroy government computer networks and data. *The American City & County, 118(6)*, p. HS6.

patches for desktops and servers. Regular system audits and log checks further support those measures.[76]

Additionally, on the local level, the Community Emergency Response Teams (CERT) program trains citizens to be better prepared to respond to emergency situations in their communities. When emergencies occur, CERT members can give critical support to first responders, provide immediate assistance to victims, and organize spontaneous volunteers at a disaster site. CERT training includes disaster preparedness, disaster fire suppression, basic disaster medical operations, and light search and rescue operations.[77] I am a proud member of the San Marcos (California) CERT. In our certification training, we covered training unit 8, which is on "Terrorism and CERT."

Sharing Information. With an increasing dependence on integrated systems, state, local, and federal agencies collectively must combat cyber attacks. Sharing information to protect systems is an important foundation for ensuring governmental continuity. States have adopted various mechanisms (such as teams, common databases, communications, and conferences) to facilitate information sharing on cyber attacks and incident reporting.[78] Information sharing has been a huge problem in the past. However, things are improving in this area of concern.

Partnership is Key. A National Association of State Chief Information Officers (NASCIO) survey noted that our information infrastructure is continually under attack. As the fight against cyberterrorism continues, the survey makes it clear that a few key improvements by the DHS—the first being a closer working relationship with state and local governments—could yield long-term benefits for state and local sectors as well as for the larger national effort.[79] Partnering is a key mechanism in resolving this issue.

---

[76] Burkhammer, L. (2006, March). The virtual enemy: Locals need more money and communication to defend against cyberterrorism. The American City & County, 121(3), pp. 32-35.

[77] Bullock, J. A., Haddow, G. D., Coppola, D., Ergin, E., Westerman, L., & Yeletaysi, S. (2006). *Introduction to Homeland Security, Second Edition.* Oxford, United Kingdom: Elsevier Butterworth-Heinemann.

[78] Anonymous (2003b, February). *The National Strategy to Secure Cyberspace.* Washington, DC: The White House.

[79] Burkhammer, L. (2006, March). The virtual enemy: Locals need more money and communication to defend against cyberterrorism. *The American City & County, 121(3)*, pp. 32-35.

Conclusion. It is only a matter of time before cyberterrorists perpetrate an attack on a city or county government. All manmade and natural disasters are local events with local units being the first to respond and the last to leave. However, the state government must be the driving force behind the efforts of defining what needs to be done, establishing a coordinated statewide strategy, and performing a comprehensive threat assessment. Sharing information to protect systems is an important foundation for ensuring governmental continuity. Partnering is a key mechanism in creating a closer working relationship among, national, state, and local governments. If these things are done, we would have a much more effective effort in combating cyberterrorism at the state and local governmental levels.

Private Industry

In this section, I identify what measures private industry is using to combat cyberterrorism.

Private Sector Responsibilities. The private sector owns and operates a large majority of our CIs/KAs. Customarily, private companies prudently engage in risk management planning. They also invest in security as a necessary component of business operations to assure customer confidence. In the current threat environment, the private sector remains the first line of defense for its own facilities. Consequently, private-sector owners and operators reassess and adjust their planning, assurance, and investment programs to accommodate the increased risk of deliberate terrorist acts. Since the events of September 11, 2001, nationwide enterprises have increased their investments in security to meet the demands of the new threat environment.[80]

Countering Cyberterrorism Rests on Private Industry. According to security experts, the U.S. government is failing to keep a close eye on the cyberterrorism threat. Rather than governments and the military, they say that the burden of watching and preparing for a computer-driven attack on U.S. CIs is falling on the private sector. Dan Vorton, executive editor of *Homeland Defense* Media in the United

---

[80] Anonymous (2003, February). *The National Strategy for the Physical Protection of Critical Infrastructures and Key Assets*. Washington, DC: The White House.

States and author of *Black Ice: The Invisible Threat of Cyberterrorism*, said, "We have an unprecedented situation where a greater part of our national security interest is in the hands of private companies, whose mission is not to protect the United States but to make money and provide shareholder value. And the government has been unwilling to increase regulation that would improve cyber security." On the contrary, Mark Rasch, former head of the U.S. Justice Department's computer crime unit, said, "There is a general danger of cyberterrorism, but there are more immediate and direct threats to the infrastructure, and if you have limited money, I would chase the other threats before cyberterrorism."[81]

Large Enterprises. Large enterprises evaluate their network security, which affect the security of the nation's CIs. Such evaluations include (1) conducting audits to ensure effectiveness and use of best practices, (2) developing continuity plans, which consider offsite staff and equipment, and (3) participating in industry-wide information sharing and best practice dissemination.[82]

Small Businesses. Small businesses help our nation secure cyberspace by securing their own connections to it. Installing firewall software and updating it regularly, maintaining current antivirus software, and regularly updating operating systems and major applications with security enhancements are actions that enterprise operators take to help secure cyberspace.[83]

Control System Security. In recent years, most companies that operate critical industrial infrastructure have invested heavily in protecting their high-level corporate information systems from cyberterrorism . . . and for good reason. Significantly, a corresponding investment in securing plant—and facility-level control systems largely has not materialized. Station equipment connects directly to industrial monitoring and control systems. A cyberterrorist attacking the control

---

[81]   Morrison, S., & Nuttall, C. (2005, November 16). US 'relying on private companies to counter cyberterrorism': But security experts are divided over the extent of the threat to the country's infrastructure. *Financial Times*, London (UK), p. 4.

[82]   Anonymous (2003, February). *The National Strategy to Secure Cyberspace*. Washington, DC: The White House.

[83]   Ibid.

system layer can cause loss of production, environmental damage, loss of intellectual property, and unsafe working conditions.[84]

Four Principles. Effectively securing CI operations from cyberterrorism requires that management observe four important principles:[85]

1. Understand that the control system layer is a vulnerable point of attack with potentially serious consequences.
2. Recognize that security tools designed for higher, corporate-level information security do not adequately address control layer security threats.
3. Plan and build a defense that handles attacks from outside as well as from within the enterprise.
4. Apply "best practices" in creating a control-level security system that will perform these five key functions: (a) monitor, (b) detect, (c) notify, (d) protect, and (e) recover.

Organizing for Cyberterrorism. Cyberterrorist threats are on the rise. Businesses could activate a proactive counterterrorism program to protect, for example, a firm's marketing activities. In this case, marketing executives can take a series of steps to minimize the potential threat and/or impact of a cyberterrorist attack. Tactical initiatives of the electronic commerce marketing team could include the following:[86]

- Develop an internal monitoring system to keep abreast of potential cyberterrorism threats.
- Maintain vigilance over system security.
- Develop action plans for the most vital threats.
- Develop a methodology for costing potential threats.
- Implement protocols for prosecuting cyberterrorists.

---

[84]   Ahern, B. M. (2003, September). Control system security in the age of cyberterrorism. *Pipeline & Gas Journal, 230(9)*, p. 12.

[85]   Ibid.

[86]   Griffith, D. A. (1999, Summer). Organizing to minimize a cyberterrorist threat: As marketing's reliance on technology increases, companies must take steps to prevent security breaches. *Marketing Management*, pp. 9-15.

Even though the idea of these initiatives may be almost a decade old, they are still current and would be effective if implemented today by companies that have not yet done anything to combat cyberterrorism.

Conclusion. Cyberterrorist threats are on the rise. However, the U.S. government is not keeping close surveillance on the cyberterrorism threat. Consequently, in the current threat environment, the private sector remains the first line of defense for its own facilities. Furthermore, since the 9/11 attacks, nationwide enterprises have increased their investments in security to meet the demands of the new threat environment. Hence, rather than governments and the military carrying the protection load, the burden of watching and preparing for a cyber attack on U.S. CIs rests on the private sector. Therefore, private industry has taken a proactive approach to combat cyberterrorism, which is a good thing.

Academia

Terrorism Education for Our Youth. There is a related requirement to develop educational programs on terrorism at all levels. Even on the primary level, students should have a basic understanding of terrorism, for they see it each day on the media and portrayed in movies and video games. By not understanding the nature of terrorism, their fears are amplified. Too often, for children and adults, it is what we don't understand of which we are most fearful. It is through education that the generalized fear of terrorism can be lessened, thereby countering the "fear multiplication" intentionally caused by acts of terrorism. Curriculum should be developed on understanding terrorism as an integral part of education in middle and high schools. It is important to develop courses that do not promote anxiety but rather give the students an understanding of a threat that is and will continue to be a reality in their own and political life.[87]

Terrorism Education for College Students. Finally, in colleges and universities, terrorism education will enable individuals to address such complex issues as reconciling security and civil liberties in a

---

[87]   Sloan, S. (2008). The evolution of terrorism as a global test of wills: A personal assessment and perspective. Memorial Institute for the Prevention of Terrorism (MIPT), 29 pp.

democratic system. But beyond that, such an education may not only help to promote interest in international affairs, but also hopefully provide the foundation for those who wish to be involved either in academia or in government as specialists on terrorism. Well-educated students with a background on terrorism are vitally needed if we are to have a new generation of individuals who will specialize in the study of a major form of international violence.[88]

New Jersey Institute of Technology. September 11, 2001, has become a symbol to the end of innocence in America and the rest of the world. "9/11" has become a synonym for terror, radical Islam, WMD's, and *Jihad*. In the aftermath of the event, many organizations have stepped into the new frontier, which can be called the counterterrorism industry. New Jersey Institute of Technology (NJIT) is one such organization. In the fall of 2006, the college initiated a new undergraduate program called "Physical and Digital Counter-Terrorism"to answer the increasing demand for counterterrorism education both from the government and the general population.[89]

Dr. Robert Statica. Dr. Robert Statica, NJIT's IT program administrator, is one of the leaders of the new program, a sequence of six courses—three in traditional terrorism and three in digital or cyberterrorism. The new program is currently open to government employees and law enforcement personnel. The IT department is also "actively working with the former Director of the FBI's NJ Joint Terrorism Task Force, retired FBI agent Stephen Foster on developing the traditional counterterrorism courses," said Statica.[90]

Cutting-edge Knowledge. The NJIT offers to companies and government agencies a practical, hands-on, and sophisticated six-course (18-credit) certificate program designed to combat digital and physical terrorism. Classes are also available at the workplace or through distance learning. "Companies or agencies may customize the program in any way they so desire. The goal is to create a new generation of counterterrorism

---

[88]   Ibid.

[89]   Shaked, S. (2006, September 12). Developing new ways to fight the war on terror. *Vector*. Retrieved May 18, 2008, from *http://www.njitvector. com/home/index.cfm?event=displayArticlePrinterFriendly&uStor y_id=baf6edc7-2178-4f61-b287-0d583bbbb6e0.*

[90]   Ibid.

experts, and we think the array of in-depth courses we've put together will fit the bill," said Robert Statica, program director. They offer cutting-edge knowledge about traditional (or physical) counterterrorism strategies, cyber security, and cyber-investigations as well as digital forensics, computer crime, and more.[91]

---

[91]   Weinstein, S. (2006, August 9). NJIT offers new certificate to fight digital and physical terrorism. NJIT Public Information Press Release. Retrieved May 18, 2008, from *http://www.njit.edu/publicinfo/press_releases/display_page.php?url=release_905.htm.*

Chapter 3
International Efforts to Combat Cyberterrorism

Robert T. Uda, MBA, MS

Combined International Efforts

International efforts to combat terrorism, much less cyberterrorism, started back around CY 2000. In a speech given on February 10, 2000, Ambassador Michael A. Sheehan, the State Department's coordinator for Counterterrorism said at the Brookings Institute, "We must 'drain the swamp'in which terrorism operates. What does that mean? We seek to limit the room [in] which terrorists have to operate, plan, move, and work. We work to show terrorists that there is no room—both physically and politically—for them to use terrorism as their means of expression . . . . When we drain the swamp—or limit the area that terrorists have to move—we expose the terrorists. Draining the swamp also means making clear to governments that they will be held accountable for controlling these areas."[92]

Coordinated International Effort. Ambassador Sheehan went further to say, "This requires a coordinated international effort to pressure those regimes, such as the Taliban in Afghanistan, which harbor terrorists, to police these swamps, expel the terrorists, and shut down areas of operation . . . . This also means putting pressure on other states—many of whom are our allies—to cut off terrorists trafficking of fighters, money, weapons, or equipment through these countries . . . .

---

[92] Sheehan, M. A. (2000, February 10). Counterterrorism chief seeks more international cooperation. Retrieved March 22, 2008, from *http://www. usembassy.it/file2000_02/alia/a0021020.htm.*

Finally, draining the swamp also means promoting a shift in public rhetoric around the world. Too often terrorism—a criminal act—is put in the light of religious expression, freedom fighting, or political statement. The international community already offers groups legitimate means of expression, and violence and terrorism are not among those legitimate forms of expression."[93]

Role of CCIPS in International Negotiations. Attorney General Janet Reno testified before the United States Senate Committee on Appropriations on February 16, 2000. Ms. Reno said, "The borderless nature of computer crime requires a large role for the Computer Crime and Intellectual Property Section (CCIPS) in international negotiations. CCIPS chairs the G-8 Subgroup on High-tech Crime, which has established a 24 hours a day/7 days a week point of contact with 15 countries for mutual assistance in computer crime. CCIPS also plays a leadership role in the Council of Europe Experts' Committee on Cybercrime, and in a new cybercrime project at the Organization of American States."[94]

The CCIPS attorney staff consists of about 40 lawyers who focus exclusively on the issues raised by computer and intellectual property crime. Section attorneys advise federal prosecutors and law enforcement agents, comment upon and propose legislation, coordinate international efforts to combat computer crime, litigate cases, and train all law enforcement groups. Other areas of expertise possessed by CCIPS attorneys include encryption, electronic privacy laws, search-and-seizure of computers, e-commerce, hacker investigations, and intellectual property crimes.[95]

DoJ Involvement at the International Level. Attorney General John Ashcroft said, "At the international level, the Department [of Justice (DoJ)] is also involved, under the leadership of the State Department, in counterterrorism coordination with other governments on a number of levels. For example, the Department represents U.S. law enforcement and prosecutorial interests in multilateral groups such as

---

[93]   Ibid.

[94]   Anonymous (2004, June 22). Computer crime and intellectual property section (CCIPS). Retrieved March 22, 2008, from *http://permanent. access.gpo.gov/lps12/ccips.html.*

[95]   Ibid.

the G-8 Counterterrorism Experts Group and in bilateral meetings with counterterrorism officials of other nations. The Department has also played a key role in developing and negotiating UN conventions relating to terrorist bombings and terrorist fund raising, both of which the United States have signed. The FBI's Legal Attachés, assigned to U.S. embassies throughout the world, assist in their response to counterterrorism issues that arise in the nations or regions they cover."[96]

Organization of American States (OAS). In remarks made to the Inter-American Committee Against Terrorism, Organization of American States, in Washington, DC, on January 28, 2002, Attorney General John Ashcroft said, "In addition, on October 29 [2001] the United States created the Foreign Terrorist Tracking Task Force. The goals of this task force are to deny entry into the United States of persons suspected of being terrorists and to locate, detain, prosecute, and deport terrorists already in the United States. We also have designed a new tamper-resistant visa and upgraded U.S. passports to prevent photo substitution. And we have intensified discussions with our friends in Canada and Mexico to improve border security. It goes without saying that, through these efforts, the United States has worked in concert with many multilateral and regional organizations. I am heartened and encouraged by the cooperation we have received from the Organization of American States, which has been supportive not only in word, but in deed."[97] It is obvious that after 9/11 had occurred, things started picking up in anti-cyber attacks.

Foreign Counterparts. Mr. John Malcolm, deputy assistant attorney general, DoJ, said, "Because cyber-attacks frequently transcend geographic boundaries, the Department's cybercrime initiatives have not been confined to the United States. It is vitally important to have foreign counterparts who are technologically capable, who are accessible and

---

[96] Ashcroft, J. (2001, May 9). The Avalon project: Statement of Attorney General John Ashcroft before the United States Committee on Appropriations Subcommittee on Commerce, Justice, and State, The Judiciary, and Related Agencies Joint Hearing on US Federal Efforts to Combat Terrorism. Retrieved March 22, 2008, from *http://www.yale.edu/lawweb/avalon/terrorism/t_0016.htm*.

[97] Ashcroft, J. (2002, January 29). Western hemisphere efforts to combat terrorism. US Department of State website. Retrieved March 22, 2008, from *http://www.state.gov/p/wha/rls/rm/7667.htm*.

responsive, and who have the necessary legal authority to cooperate with us and assist in our investigations and prosecutions in the event of a trans-border cyber incident."[98]

24/7 Network. John Malcolm further said, "We are working hard to build strong relationships with foreign counterparts so that the framework will be in place to quickly respond to cybercrimes, including large-scale cyber incidents. For example, CCIPS chairs (and has chaired since its inception in 1997) the G-8 Subgroup on High-tech Crime. One of the most significant achievements of this Subgroup is the creation of the '24/7 Network,' which allows law enforcement in the participating countries to reach out '24 hours a day, 7 days a week' to counterparts in other countries for rapid assistance in investigating computer crime and preserving electronic evidence. Often, cyber-criminals can be identified only if evidence of their conduct is preserved within minutes, a time-frame that is way too short for us to rely on traditional international assistance options."[99]

A Chain is as Strong as Its Weakest Link. Malcolm further said, "Currently, 35 countries participate in the 24/7 Network. This network has been used successfully in many instances to investigate threats and other crimes in a number of countries, including the United States. Because terrorists operate throughout the world, it is critical that we continue our efforts to expand the Network in order to ensure that our law enforcement capabilities are coexistensive (sic). When it comes to combating cybercrime across international boundaries, the chain is truly only as strong as its weakest link."[100]

Pacific Rim Countries. Looking at the Pacific Rim countries, Singapore actively participated in counterterrorism efforts through various international forums, including the ASEAN Regional Forum. In March 2005, Singapore and the United States co-hosted an ASEAN Regional Forum confidence-building measure conference on maritime

---

[98]   Malcolm, J. (2004, February 24). Virtual threat, real terror: Cyberterrorism in the 21st century. Testimony Before the United States Senate Committee on the Judiciary. Retrieved March 22, 2008, from *http://www.globalsecurity/ org/security/library/congress/2004_h/040224-malcolm.htm.*

[99]   Ibid.

[100]   Ibid.

security. In June 2005, Singapore hosted the first Asia-Middle East Dialogue, with government and non-government representatives from 50 countries. Senior Minister Goh Chok Tong proposed the dialogue as a "platform for progressive Muslim voices."[101] Contrary to what some left-leaning liberal politicians say, the United States is cooperating and gathering support from other nations in fighting terrorism.

Council of Europe Parliamentary Assembly. The Council of Europe Parliamentary Assembly notes that the fight against cybercrime requires urgent international cooperation among governments, the private sector, and non-governmental organizations (NGOs) as cybercriminals rely on their ability to operate across borders and to exploit differences in national laws. The lack of cooperation by the member states exposes them to considerable danger. In this context, the Parliamentary Assembly welcomes the various initiatives taken in order to enhance international cooperation and coordination in the fight against cybercrime, the 24/7 points of contact, and the "Check the Web" program. Additionally, it strongly encourages member states to continue to reinforce their efforts, to strengthen international cooperation, and to support concrete, coordinated measures for more efficient protection. Furthermore, the relevant laws need to be standardized, or at least be compatible with one another, to permit the required level of international cooperation.[102] Europe is working together to fight cybercrime and cyberterrorism.

Summary. In summary, as can be seen from the above discussion, the United States is making combined international efforts to combat cyberterrorism. This approach is an important and sure way to beat cybercriminals and cyberterrorists.

---

[101] Anonymous (2006, April 26). Singapore—Excerpt DoS Country Report on Terrorism. Retrieved March 22, 2008, from *http://asiasecurity.org/ issue-papers/singapore-excerpt-dos-country-report-on-terrorism-26-april-2006.*

[102] Anonymous (2007, June 28). How to prevent cybercrime against state institutions in member and observer states? Parliamentary Assembly Resolution 1565 (2007). Retrieved March 22, 2008, from *http://assembly. coe/int/Documents/AdoptedText/ta07/ERES1565.htm.*

Cyber Protection Measures at the International Level

In this section, I identify cyber protection measures being used at the International level that I believe could be effective here in the United States. Other countries of the world use cyber protection measures that the United States could adopt. I address these measures in the ensuing paragraphs including my reasoning as to why they could also be effective in the United States.

China. China maintains some of the world's tightest government restrictions on Internet use. These restrictions make many observers skeptical that hacker gangs could operate within China without government approval or acquiescence.[103] This approach would certainly reduce drastically the effects of hackers, hactivists, and foreign government cyber-attackers against our U.S. cyber-systems. However, unless we were engaged in an all-out war, we would experience a huge backlash from civil liberties organizations such as the American Civil Liberties Union (ACLU). The USA PATRIOT Act goes far enough for the United States to fight terrorists but too far for the ACLU and other such organizations.

Germany. On February 27, 2008, Germany's top court ruled that domestic security services may monitor the computers of suspected criminals or terrorists. However, they may do so only if they possess evidence showing the suspects to be dangerous. The decision by the Constitutional Court opens the door to a new federal law. Chancellor Angela Merkel's conservative Christian Democrats have long advocated this new law, which would allow online surveillance but only under strict conditions.[104] This measure could also be effective in the United States. However, like what is happening in China with hackers and the Internet, there would be a huge outcry from civil liberties organizations unless we were engaged in an all-out war. Like

---

[103] *The Washington Times* (2008, March 24). Cyber-attacks on Tibet groups tied to China. *National Security Research & Education Programs.* Retrieved April 5, 2008, from *http://homelandsecurity.osu.edu/focusareas/cyberterrorism.html.*

[104] Reuters (2008, February 28). German court permits limited cyber-monitoring. *NYTimes.com.* Retrieved April 5, 2008, from *http://homelandsecurity.osu.edu/focusareas/cyberterrorism.html.*

above, The USA PATRIOT Act goes far enough to fight terrorists but too far for ACLU's comfort.

Singapore. In February 2006, Singapore announced it would spend $23 million over three years to enhance computer security and combat cyberterrorism. Singapore's Parliament amended the Money-changing and Remittance Businesses Act in August 2006. This act strengthens the government's ability to combat money laundering and terrorist finance-related activities in the money-changing and remittance sector.[105] The United States covers these areas well with the Bank Secrecy Act, Money Laundering and Financial Crimes Strategy Act of 1998, The USA PATRIOT Act, The Money Laundering Regulations 2003, and the 2003 National Money Laundering Strategy.

Hong Kong. Many activities within the region are of considerable interest, particularly the successful efforts in Hong Kong to establish tertiary diploma-level training in computer investigations and computer forensics through a partnership relationship between the Hong Kong Police and the Hong Kong University of Science and Technology.[106] The United States covers this area fairly well with the many private schools and government training centers that provide degrees, certifications, and special education/training in computer investigations and forensics.

IACIS. The most prominent training organization in the law-enforcement-only training area is the International Association of Computer Investigation Specialists (IACIS). Other law enforcement agencies, both on national and local levels, offer courses on a more sporadic basis. These classes not only cover the technical aspects of computer forensics but also concentrate on criminal law issues. Vendors who sell computer forensic tools offer vendor-specific training courses. These courses generally cover in detail the features that make their tools powerful.[107]

---

[105] Anonymous (2006, April 26). Singapore—excerpt DoS country report on terrorism. *Southeast Asia—Security Information Resource*. Retrieved March 22, 2008, from *http://asiasecurity.org/issue-papers/singapore-excerpt-dos-country-report-on-terrorism-26-april-2006*.

[106] Anonymous (2002, February). Technology research: Interpol Asia-South Pacific working party on IT crime. *Australasian Center for Policing Research* (ACPR) *Bulletin, Number 11*.

[107] Kuchta, K. J. (2001, November-December). Learning the Computer Forensic Way. *Information Systems Security, 10(5)*, 29-35. Retrieved April

Computer forensic overview training classes focus on the academic theory of computer forensics. Mostly professional organizations and associations offer a number of these courses to the private sector. Do not expect to attend one of these courses to become a computer forensic expert. Practitioner hands-on training courses are geared toward the computer forensic practitioner. The duration of most of these courses is five days. These training classes may consist of upwards of 50 percent hands-on exercises. A good computer forensics professional possesses three different types of skill sets in one person, i.e., help desk specialist, programmer, and investigator.[108]

Organization of American States (OAS). In response to the international terrorist threat, the OAS took important steps to help prevent terrorism from establishing itself in the Western Hemisphere. Within 4-1/2 months of the September 11 attacks, the Inter-American Committee Against Terrorism (CICTE) had focused correctly on practical results. Attorney General John Ashcroft commended the committee for promoting concrete action among member states in three critical areas:[109]

1. Tightening border controls against those who would enter a country to commit terrorism
2. Establishing more effective networks and mechanisms to track and intercept the financing of terrorists
3. Sharing each others' experiences through training and joint exercises

In this regard, Ashcroft underscored the important role played by Argentina, as vice-chair of CICTE. He also commended the leadership shown by El Salvador, Peru, and Colombia in chairing the key working

---

7, 2008, from *http://firstsearch.oclc.org.proxy1.ncu.edu/images/WSPL/wsppdf1/HTML/06470/MAYZ6/DSV.HTM.*

[108] Ibid.

[109] Ashcroft, J. (2002, January 28). Western hemisphere efforts to combat terrorism. US Department of State. Remarks made to the Inter-American Committee Against Terrorism (CICTE), Organization of American States (OAS). Retrieved March 22, 2008, from *http://www.state.gov/p/wha/rls/rm/7667.htm.*

groups within the organization.[110] The United States could benefit by controlling its borders as some of our Latin American countries do . . . with guns and bullets.

Council of Europe. The Assembly noted that the fight against cybercrime requires urgent international cooperation among governments, the private sector, and nongovernmental organizations. This is because cybercriminals rely on their ability to operate across borders and to exploit differences in national law. Lack of cooperation by member states exposes them to considerable danger. Furthermore, the relevant laws need to be standardized, or at least made compatible with one another, to permit the required level of international cooperation.[111] More cooperation and standardized/compatible laws with other nation states would make the United States more effective in dealing with cybercrime.

European Union (EU). Dr. Norman Neureiter made the point that the U.S. research community addressing critical infrastructure protection (CIP) issues is totally inadequate. He said that he was sure it was likewise in the EU. While pockets of significant research exist, the number of people *in toto* is not as large as it needs to be. That is where education comes in. We need more training in cyber security. This shortage of people makes international collaboration an even greater imperative. We need to combine our small numbers.[112]

Cyber systems are global, and we all face the same terrorism problems. Hence, CIP and cyber security comprise issues which demand international science and technology (S&T) collaboration and coordination. We cannot solve the global challenges in one country

---

[110] Ibid.

[111] Anonymous (2007, June 28). How to prevent cybercrime against state institutions in member and observer states? Council of Europe, Parliamentary Assembly, Resolution 1565. Retrieved March 22, 2008, from *http://assembly.coe.int/Documents/AdoptedText/ta07/ERES1565. htm.*

[112] Neureiter, N. P. (2001, December 3). Critical infrastructure protection: International S&T cooperation after September 11. Dr. Norman P. Neureiter's remarks to the IST 2001 Technologies Serving People Conference in Dusseldorf, Germany. Department of State. Retrieved April 2, 2008, from *http://www.state.gov/g/stas/8580.htm.*

only. We need to work on them together. Three years ago, the United States and EU recognized that fighting it alone is simply not an option. However, cooperation among governments alone is not enough. We also need partnerships with industry.[113] Working together, education and training, collaboration and coordination, and industry partnerships must increase in the United States for us to be more effective in combating cyberterrorism.

The International Framework. A key benefit of a global solution is that it can approach the problem of network security from an international political perspective. Thus, the possible range of solutions and options available to decision-makers is potentially much wider than nation-centric or local efforts to mitigate attacks. Reaching a global solution requires a proper framework. This framework should consist of the following:[114]

- *Model.* Agreement on a common critical infrastructure model
- *Trust and Confidence.* Trust and confidence building among the participating countries and organizations
- *Participation.* Active government participation at the policy level
- *Approach.* A collective security approach

Such an international framework would allow nation states and the UN, EU, OAS, and other organizations to be legally tough on terrorism.

International Cooperation. Ambassador Michael A. Sheehan, coordinator for counterterrorism, spoke at the Brookings Institution on February 10, 2000. He said that we seek to limit the room in which terrorists can operate, plan, move, and work by outlining a few of our key activities:[115]

---

[113] Ibid.

[114] Bryen, S. (2002, May 20). A collective security approach to protecting the global critical infrastructure. International Telecommunication Union (ITU) Workshop on Creating Trust in Critical Network Infrastructures, Document: CNI/09, Seoul, Republic of Korea, 20-22 May 2002.

[115] Sheehan, M. A. (2000, February 10). Counterterrorism chief seeks more international cooperation. Post-Millennium Terrorism Review speech given

- *State Sponsors.* We pressure state sponsors, thereby isolating them from the international community
- *FTOs.* We criminalize terrorism through the process of designating Foreign Terrorist Organizations (FTOs)
- *Separate Actor from Action.* We depoliticize the message of terrorism through public statements separating actor from action
- *Consensus.* We build international consensus for zero tolerance by working with our G-8 and EU partners
- *International Framework.* We support the construction of an international legal framework to allow states and the UN, EU, OAS, and other organizations to be legally tough on terrorism
- *Work Bilaterally.* We work bilaterally to arrest, disrupt, and expel terrorists
- *Training.* Finally, we bolster the capacity of those countries that need it to fight terrorism through our international training programs, which are run by state, law enforcement, and intelligence agencies

If the United States will do these things that we are not currently doing or are not doing well, we should do them to be more effective in fighting the Global War on Terrorism (GWOT).

In summary, I have identified a few cyber protection measures being used at the international level by other countries. I believe that some of these measures could be effective applied here in the United States. If the United States will add these protective measures to our present repertoire of tools, we will be even more effective in combating cyberterrorism.

---

by Ambassador Michael A. Sheehan, coordinator for counterterrorism, Brookings Institution. Retrieved March 22, 2008, from *http://www. usembassy.it/file2000_02/alia/a0021020.htm.*

Chapter 4
U.S. Policy to Prevent a Cyber Attack

Robert T. Uda, MBA, MS

T echnology has outpaced policy in cyberspace. Today's cyber criminals are technologically agile, perceptive, and evasive. To beat them, there must a unified effort between government and the private sector. There are no rules about what government or private entities can do if terrorists attack their networks.

The United States is already in a cyber war. Yet current policies prevent the United States from pursuing cyber threats based in and originating from foreign countries. As is usually the case, we create policies to hamper ourselves, not hamper the enemy. Obviously, current U.S. cyberwarfare strategy is dysfunctional. To compound matters, incompatible U.S. cyber forces do not communicate with one another, thereby, resulting in a disjointed effort. Hence, it may take a cyber version of the 2001 terrorist attacks for the country to realize it must re-examine its approach to cyberwarfare.

In the *National Strategy for Homeland Security*, the Homeland Security Council states that many of our nation's essential and emergency services, as well as our critical infrastructure, rely on the uninterrupted use of the Internet and the communications, data, monitoring, and control systems that comprise our cyber infrastructure. A cyber attack could be debilitating to our highly interdependent critical infrastructure and key resources (CI/KR) and ultimately to our economy and national security.[116]

As a nation, our policies to prevent cyberterrorism need to help shore up our strengths and help convert our weaknesses into strengths.

---

[116] Anonymous (2007, October). *National Strategy for Homeland Security*. Washington, DC: Homeland Security Council.

This chapter presents an evaluation of the strengths and weaknesses of U.S. policy to prevent a cyber attack.

Computer Security Effects on Critical Infrastructure

Our Nation's Critical Infrastructure. The arsenal of modern weapons that terrorists might someday use to disrupt power grids, gas lines, and other parts of the nation's critical infrastructure includes conventional weapons as well as bits and bytes—in other words, cyberterror attacks. The cyber threat to the electricity we use and the water we drink is real, experts say, but there's no need to panic—at least not yet.[117] The trouble is, we never panic *before* a major terrorist event, but we do panic *after* a major terrorist event. Take 9/11 for example, Congress finally did something by passing legislation and allocating sufficient funding to fight the Global War on Terrorism (GWOT). Now that nearly seven years have passed since 9/11, everyone is falling back into a deep sleep. Some people do not believe we face a cyberterror threat. However, just wait until it happens, then, Congress will take some significant actions to counter that threat. Let us hope it will not be too little too late.

Threat to Our Water Supplies. Regarding the cyber threat to the water we drink, automated water supply control systems have long been a subject of concern from U.S. infrastructure protection specialists. A 1997 report by the Clinton Administration's Presidential Commission on Critical Infrastructure stated, "Cyber vulnerabilities include the increasing reliance on SCADA [Supervisory Control and Data Acquisition] systems for control of the flow and pressure of water supplies."[118] Hence, the threat is real and anticipated.

Securing Cyberspace. According to an estimate, the cost to our economy from attacks on our information systems has grown by 400 percent within four years, but it still has limits. In one day, however, that could drastically change. Every day, somewhere in America, an

---

[117] Blau, J. (2004, November 29). The battle against cyberterror: The race is on to harden the nation's critical infrastructure before cyberterrorists gain enough skills to launch attacks. *Network World, 21(48)*, p. 49.

[118] Swichtenberg, B. (2002, March). FBI issues water supply cyberterror warning. *Water Engineering & Management, 149(3)*, p. 7.

individual company or a home computer user suffers what, for them, are significantly damaging or catastrophic losses from cyber attacks. The ingredients are present for that kind of damage to occur on a national level, i.e., to our national networks and the systems they run upon and of which the nation depends. Our potential enemies have the intent. Their tools of destruction are broadly available. Additionally, the vulnerabilities of our systems are myriad and well known. In cyberspace, a single act can inflict damage in multiple locations simultaneously without the attacker ever physically needing to enter the United States.[119] In January 2008, I experienced a personal computer crash, which totally locked up my computer. I suspect that a virus brought in from an email attachment destroyed the hard drive. As discussed earlier in this paragraph, the catastrophic loss of data I had experienced was damaging to me.

Cyber Incidents. A cyber-related incident of national significance may take many forms: (1) an organized cyber attack, (2) an uncontrolled exploit such as a virus or worm, (3) a natural disaster with significant cyber consequences, (4) or other incidents capable of causing extensive damage to critical infrastructure or key assets. Large-scale cyber incidents may overwhelm government and private sector resources by disrupting the Internet and/or taxing critical infrastructure information systems. Complications from disruptions of this magnitude may threaten lives, property, the economy, and national security. Rapid identification, information exchange, investigation, and coordinated response and remediation often can mitigate the damage caused by this type of malicious cyberspace activity.[120]

A Real-life Example. I work at BAE Systems, Inc., one of the largest defense firms in the world. BAE constantly upgrades its information technology (IT) systems to counter hackers and trained adversaries from foreign countries. The BAE Systems command, control, computing, & intelligence (C3I) business area located in San Diego, California, possesses a business continuity and emergency

[119] Anonymous (2002, July). *National Strategy for Homeland Security*. Washington, DC: Office of Homeland Security.

[120] Bullock, J. A., Haddow, G. D., Coppola, D., Ergin, E., Westerman, L., & Yeletaysi, S. (2006). *Introduction to Homeland Security, Second Edition*. Oxford, United Kingdom: Elsevier Butterworth-Heinemann.

response plan that covers emergency response, crisis management, and business recovery. This company is prepared to deal with terrorist action, HAZMAT incidents, and natural disasters . . . among other threats. Recently, the company successfully employed/deployed its emergency response plan and emergency response team during the wildfires in San Diego County. All response activities went according to plan.

Tighter Physical Security Promulgate Cyber Attacks? Clay Wilson, a technology and national security specialist at the Congressional Research Service (CRS), said that tighter physical security measures in the United States might encourage terrorist groups in the future to explore cyber attacks. Terrorists are not yet mobilizing to carry out extensive cyber attacks, Wilson said. Although the possibility remains, it is extremely difficult to know if they will take their current cyber activity to the next level to inflict physical harm, said Wilson.[121] We must not remain complacent or bury our collective heads in the sand, but we should be constantly vigilant. Hence, when the attack comes, we will be mentally prepared for it.

Legal Requirements of Computer and Cyber Security

Electronic Communications Privacy Act. The Electronic Communications Privacy Act of 1986 (ECPA), without authorization, makes it illegal to intentionally access a facility providing electronic communications services, or to intentionally exceed the authorization of access to such a facility. The bill intended to protect the privacy of high-tech communications such as electronic mail, video conference calls, conversations on cellular telephones, and computer-to-computer transmission.[122] However, over the five-year period from 1998-2002, the number of cyber attacks on telephone companies (telcos) and their networks had increased significantly. For example, Carnegie Mellon University's computer emergency response team (CERT), based in the

---

[121]   Wagner, B. (2007, July). Electronic jihad: Experts downplay imminent threat of cyberterrorism. *National Defense*, pp 34-36.

[122]   Sweet, K. M. (2006). *Transportation and Cargo Security: Threats and Solutions.* Upper Saddle River, New Jersey: PEARSON Prentice Hall.

United States, reported 3,700 attacks on telco networks in 1998 while in 2002, it was set to increase to a staggering 110,000 attacks.[123] Obviously, the ECPA minimally affected cyber attackers.

Computer Security Act. Subsequently, the Computer Security Act, passed a year later in 1987, was the first major legislation relating to information security. This bill provides "for a computer standards program within the National Bureau of Standards [now called NIST, National Institute of Standards & Technology] to provide for government-wide computer security, to provide for the training in security matters of persons who are involved in the management operation and use of federal computer systems, and for other purposes."[124],[125] However, just 15 years later in mid-2002, to the question of how prepared are U.S. companies for a cyberterrorist attack, Michael Vatis, director of the Institute for Security Technology Studies (ISTS) at Dartmouth College said, "As a general matter, companies, government agencies, and academia are inadequately prepared. Too little attention is paid to security; too few resources are devoted to it."[126] Mind you, this was almost a year after 9/11!

Clinger-Cohen Act. In 1996, the Clinger-Cohen Act created the position of Chief Information Officer (CIO) within government agencies to ensure that they properly acquire and manage information systems. Further, based upon NIST-developed standards, the act called for the Secretary of Commerce to "promulgate standards and guidelines pertaining to federal computer systems." Additionally, "the secretary shall make such standards compulsory and binding to the extent to which the secretary determines necessary to improve the efficiency of operation or security and privacy of federal computer

---

[123] Ashenden, D. (2003, January). Protect and survive. *Telecommunications International, 37(1)*, p. 29.

[124] Thomas (1987). Computer Security Act of 1987. Retrieved (n.d.) from *http://thomas.loc.gov/cgi-bin/bdque.*

[125] Cordesman, A. H. (2002). *Cyber-threats, Information Warfare, and Critical Infrastructure Protection: Defending the US Homeland.* Westport, Connecticut: Praeger Publishers.

[126] Gaudin, S. (2002, July 19). Security expert: US companies unprepared for cyberterror. *IT Management.* Retrieved November 25, 2002, from *http://itmanagement.earthweb.com/secu/print.php/1429851.*

systems."[127],[128] To the question that if there is one thing you could tell these CIOs and CSOs (chief security officers) to do, Michael Vatis said, "There's no one thing, but one message is that senior management needs to make security a priority. CEOs and boards of directors need to pay attention to security and make sure resources are devoted to it."[129] Company CEOs and boards of directors should pay heed to this clarion call to proactive action.

Availability of Law. The principle of *nullum crimen sine lege* is fundamental to most legal systems. Under this principle, any behavior no matter how harmful, the legal system cannot prosecute unless that behavior is formally prohibited by law. For example, the Philippine courts could not prosecute the person who released the I LOVE YOU virus in May 2000 because, at the time, there was no law in the country that prohibited the release of malicious code. The history of criminal law in many countries is replete with examples of newly created laws to cope with new forms of undesirable behavior. The advent of digital technology necessitates numerous legislative activities to this end. Indeed, very soon after the release of the I LOVE YOU virus, the Philippine government introduced legislation to criminalize virus dissemination.[130] However, as usual, governments and politicians are nearly always reactive . . . seldom if ever proactive.

Difficulties and Needs of a Computer Forensics Program

Legal Evidence. By most estimates, 90 percent of legal evidence resides in computer systems rather than on paper. Cases involving

---

[127] Anonymous (2000). Information Technology Management Reform Act of 1996. Retrieved July 5, 2000, from *http://www.rdc.noaa.gov/~irm/div-e. htm.*

[128] Cordesman, A. H. (2002). *Cyber-threats, Information Warfare, and Critical Infrastructure Protection: Defending the US Homeland.* Westport, Connecticut: Praeger Publishers.

[129] Gaudin, S. (2002, July 19). Security expert: US companies unprepared for cyberterror. *IT Management.* Retrieved November 25, 2002, from *http:// itmanagement.earthweb.com/secu/print.php/1429851.*

[130] Grobsky, P. (2007, October 13). Requirements of prosecution services to deal with cybercrime. *Crime Law Soc Change, 47,* 201-223.

corporate trade secrets, personal and commercial disputes, employment discrimination, misdemeanor and felony crimes, and personal injury can be won or lost solely with the introduction of recovered e-mail messages and other electronic files and records.[131] Hence, it is important that companies preserve their computer-stored data for the required duration of time.

Trained Computer Forensics Specialist. Computer forensics analysis is becoming increasingly useful to businesses. Computers can contain evidence in many types of human resources proceedings including sexual harassment suits, allegations of discrimination, and wrongful termination claims. Specialists can find evidence in electronic mail systems, network servers, and individual employee's computers. However, due to the ease with which perpetrators can manipulate computer data, the court could likely reject the data as evidence if a trained computer forensics specialist does not perform the search/analysis.[132]

National Computer Forensic Institute. In March 2007, the U.S. Department of Homeland Security (DHS) opened the National Computer Forensic Institute in Hoover, Alabama, to assist in the field of computer forensics and digital evidence analysis. The U.S. Secret Service is developing this institute along with partial funding from the department's National Cyber Security Division. It serves as a national cybercrimes training facility where they offer training and equipment to state/local police officers and prosecutors/judges.[133]

Difficulties and Needs to Effect Results in Public and Private Venues

Private Sector Role in Cybercrime Control. Effective and efficient control of cybercrime requires more than cooperation among law

---

[131] Philpott, D. (n.d.). Special report: Computer forensics and cyber security. *Homeland Defense Journal*, 12 pp.

[132] Anonymous (n.d.). Overview of Computer Forensics Technology—Part I. Retrieved March 19, 2008, from *www.charlesriver.com/resrcs/chapters/1584503890_1stChap.pdf*.

[133] Philpott, D. (n.d.). Special report: Computer forensics and cyber security. *Homeland Defense Journal*, 12 pp.

enforcement agencies. We cannot understate the role of communications and IT industries in designing crime-resistant products that facilitate detection and investigation. Furthermore, actual collaboration of private sector organizations with public law enforcement agencies is already a fact-of-life in some countries. For example, with 680,000 subscribers, Pacific Century Cyberworks (PCCW) is Hong Kong's largest Internet service provider (ISP) with a dedicated department that helps Hong Kong Police (HKP) conduct joint investigations.[134]

FBI Reaches Out to Private Sector. For a decade, federal law enforcement officials preached the gospel of private-sector cooperation. The need has long been obvious, but Dan Larkin, Unit Chief of the FBI's Internet Crime Complaint Center, said the government is getting serious about the effort. "We need to go after these partnerships more aggressively," said Larkin. The stakes in the game of cat and mouse between law enforcement and cyber criminals are getting higher. Spam and cybercrime are really about the money. There are people making a lot of money out there," he said.[135]

Government and Industry on the Defensive. In recent years, computer crimes have increased rapidly and have overtaken the ability of the government and private sector to protect their systems fully. "We are constantly in the reactive mode," says Jerry Dixson, director of the National Cyber Security Division at the DHS. Long gone are the days of young whiz kids hacking into computers to fulfill curiosity or to prove their computer prowess. Today's cyber criminals are technologically agile, perceptive, and evasive. "Software we found through investigations has gotten really sophisticated. It almost requires a PhD," Dixson says.[136]

Central Communications Platform. *The bottom line need is this:* Many Wall Street IT executives say that the most effective way to deal with the twin threats posed by hackers and cyberterrorists is to form, using a private website, a unified effort between the government and

---

[134]   Grobsky, P. (2007, October 13). Requirements of prosecution services to deal with cybercrime. *Crime Law Soc Change, 47*, 201-223.

[135]   Philpott, D. (n.d.). Special report: Computer forensics and cyber security. *Homeland Defense Journal*, 12 pp.

[136]   Wagner, B. (2007, October). Electronic attackers: Computer crimes keep government and industry on the defensive. *National Defense*, pp 24-26.

private sector to create a central communications platform to alert and disseminate information about such attacks.[137] The State of New Jersey has NJ-Alert, which is an emergency alert system that sends messages to subscribers through their emails and cell phones. During times of emergency (such as natural disasters and terrorist threats), this system allows agencies to contact the public quickly. Alerting the public of an emergency and telling them how to respond to it properly provide the greatest amount of public safety.[138]

U.S. Policy to Prevent a Cyber Attack

National Strategy Document and Cyber Security Plan. In order to secure our cyber infrastructure against manmade and natural threats, our federal, state, and local governments, along with the private sector, are working together to prevent damage to and unauthorized use and exploitation of our cyber systems. We also are enhancing our ability and procedures to respond to an attack or major cyber incident. The *National Strategy to Secure Cyberspace* and the National Institute of Public Policy's Cross-Sector Cyber Security plan are guiding the U.S.'s efforts.[139]

The *National Strategy to Secure Cyberspace* document defines how the DHS ensures information security and protects cyberspace. This document covers three strategic objectives to secure cyberspace as follows:[140]

1. Prevent cyber attacks against America's critical infrastructures.
2. Reduce national vulnerability to cyber attacks.
3. Minimize damage and recovery time from cyber attacks that do occur.

---

[137] Mearian, L. (2002, March 4). Wall Street seeks cyberterror defenses. *Computerworld, 36(1)*, p. 7.

[138] NJ-Alert (2008). New Jersey Alert System website. Retrieved March 19, 2008, from *http://www.njalert.org/*.

[139] Anonymous (2007, October). *National Strategy for Homeland Security*. Washington, DC: Homeland Security Council.

[140] Bullock, J. A., Haddow, G. D., Coppola, D., Ergin, E., Westerman, L., & Yeletaysi, S. (2006). *Introduction to Homeland Security, Second Edition*. Oxford, United Kingdom: Elsevier Butterworth-Heinemann.

Telecommunications Program and Cyber security Policy. The United States needs robust telecommunication services to manage national defense, disaster response, and other emergency services. The Program on Telecommunications and Cyber security Policy is concerned with issues relating to (1) protecting and maintaining a resilient telecommunications industry and infrastructure, (2) protecting against or responding to cyber attacks on telecommunication, network, or computer infrastructure, and (3) regulatory policy relating to telecommunications infrastructure and spectrum management, ownership, and control.[141]

Cyber Safehavens. The Internet provides an inexpensive, anonymous, geographically unbounded, and largely unregulated virtual haven for terrorists. Our enemies use the Internet to develop and disseminate propaganda, recruit new members, raise and transfer funds, train members on weapons use and tactics, and plan operations. Terrorist organizations can use virtual safehavens based anywhere in the world regardless of their members' or operatives'location. Internet use, however, also creates opportunities for us to exploit. To counter terrorists using the Internet as a virtual sanctuary, we discredit terrorist propaganda by promoting truthful and peaceful messages. We ultimately seek to deny the Internet to the terrorists as an effective safehaven for their propagandizing, proselytizing, recruiting, fundraising, training, and operational planning.[142]

Startling Security Gaps. America's government and defense contractors have been victimized by an unprecedented rash of cyber attacks over 2006-2007, forcing the United States to launch a new operation to fight off intrusions. In an article headlined "The New E-Spionage Threat," *Business Week* disclosed that a probe of the attacks on sensitive computer networks uncovered "startling security gaps." "It's espionage on a massive scale," Paul Kurtz, a former high-ranking national security official, told the magazine. In 2007, government agencies reported nearly 13,000 "cyber security incidents" to the DHS, three times the number from 2005. Incidents involving the military's

---

[141] Anonymous (n.d.). The Program on Telecommunications and Cyber security Policy. The Global Information society Project. Retrieved March 20, 2008, from *http://www.telecom-program.org/*.

[142] Anonymous (2006, September). *National Strategy for Combating Terrorism.* Washington, DC.

networks rose 55 percent in 2007. Private targets, such as defense contractors, are also vulnerable and information gleaned from their computers could pose a serious security risk.[143]

Strengths and Weaknesses of the U.S. Policy

Currently No Rules. The United States is working to prevent attacks on military, government, and private computer networks, but any aggressive response raises legal, civil rights, and policy questions that we should address, said a U.S. military adviser in November 2007. At the moment, "there are no rules" about what government or private entities can do if terrorists attack their networks, said Andrew Palowitch, chief technology officer for Science Applications International Corporation's Intelligence and Security Group and senior adviser to the Pentagon. Given more than 37,000 attempted breaches of government and private programs and 80,000 attacks on military networks in fiscal 2007 alone, some would argue that the United States was already in a "cyber war," he said in a speech at Georgetown University in Washington, DC. However, he said he was not speaking for the Pentagon.[144]

Current Policies Prevent United States. Senior military officials have spoken out recently on U.S. cyber strategy saying the country urgently needs to develop new policies and procedures for fighting in the cyber domain. Current U.S. cyberwarfare strategy is dysfunctional, said Gen. James Cartwright, commander of the Strategic Command (Stratcom), in a speech at the Air Warfare Symposium in Orlando, Florida, in February 2007. Offensive, defensive, and reconnaissance efforts among U.S. cyber forces are incompatible and do not communicate with one another, thereby, resulting in a disjointed effort, Cartwright said. Gen. Roland Keys, commander of Air Combat Command, told reporters at the conference that current policies prevent the United States from pursuing cyber threats based in foreign countries. Technology has

---

[143] Anonymous (2008, May 25). US battling 'new espionage threat'—cyber attacks. *Newsmax.com.* Retrieved May 25, 2008, from *http://news. newsmax.com/?Z6OD.ZdVAp2FmyHW-TtemhQZzXlztfR1Z.*

[144] Shalal-Esa, A. (2007, November 27). US working to respond to growing cyber attacks. Reuters. Retrieved March 20, 2008, from *http://www. reuters.com/article/companyNewsAndPR/idUSN2753754120071127.*

outpaced policy in cyberspace, he said. The United States should take added aggressive measures against foreign hackers and websites that help others attack government systems, Keys said. It may take a cyber version of the 2001 terrorist attacks for the country to realize it must re-examine its approach to cyberwarfare, he added.[145]

Workplace Policies Possess both Good and Bad Traits. Although flexible work schedule and telecommuting policies tend to increase the productivity and satisfaction of employees,[146] these labor policies can place employers at risk as related to cyber attacks. The policy of *flexi-time* can increase the risk of cyber attacks because employees work nontraditional hours. These hours may give dishonest or disgruntled employees a better opportunity to steal, modify, or view secure information. The policy of *telecommuting* can also compromise cyber security because it offers hackers another way into a company's system. In addition, it is unlikely that home network connections are as secure as work connections.[147]

Conclusion

The Electronic Communications Privacy Act does not curtail cyber attackers. Generally, companies, government agencies, and academia are inadequately prepared. We pay too little attention to security. We devote too few resources to it. Management needs to make security a priority. CEOs and boards of directors need to pay attention to security and make sure resources are devoted to it.

Under the principle of *nullum crimen sine lege*, the legal system cannot prosecute a perpetrator unless the law prohibits his/her behavior (no matter how harmful it may be). Hence, the Philippine government

---

[145] Rogin, J. (2007, February 13). Cyber officials: Chinese hackers attack 'anything and everything.' *FCW.com*. Retrieved March 20, 2008, from *http://www.fcw.com/online/news/97658-1.html*.

[146] Kundu, K. (1999, November 23). Telecommuting: Work is virtually something you do, not somewhere you go. *Future Trends*. Employment Policy Foundation.

[147] Jacknowitz, A. (n.d.). Cyber attacks! Trends in US corporations. The Business Forum. Retrieved March 20, 2008, from *http://www.bizforum. org/whitepapers/rand001.htm*.

did not prosecute the perpetrator of the "I LOVE YOU" virus because no law existed that prohibited the release of malicious code.

It is important that companies preserve their computer-stored data for the required duration of time. Due to the ease with which perpetrators can manipulate computer data, the court could likely reject the data as evidence if a trained computer forensics specialist does not perform the search/analysis. It is important that ISPs help law enforcement to conduct joint investigations. The stakes in the "cat and mouse" game is getting higher. Spam and cybercrime are really about the money. There are people making a lot of money out there. To curtail cybercrimes, there must be a unified effort between government and the private sector.

Chapter 5
Effectiveness of U.S. Infrastructure Programs

Robert T. Uda, MBA, MS

This chapter covers an evaluation of the effectiveness of U.S. infrastructure programs. I analyzed the admissibility and acquisition of evidence related to computer crime. I also critiqued the problem of cyberterrorism and identify international and local efforts to combat it. This chapter covers the following topics:

- Analysis of the Acquisition of Computer Crime Evidence
- Analysis of the Admissibility of Computer Crime Evidence
- A Critique of the Cyberterrorism Problem
- International and Local Efforts to Combat Cyberterrorism
- An Evaluation of the Effectiveness of U.S. Infrastructure Programs

Analysis of the Acquisition of Computer Crime Evidence

In this section, I analyzed the acquisition of computer crime evidence and the necessary requirements and care that must be taken while gathering this cyber-criminal evidence.

Electronic Evidence: Ever Present, Not Easily Removed. Anyone and everyone—even a sophisticated hacker—using a computer for any kind of activity leaves behind potential electronic evidence or, if you may, a cyber trail. We shop, research, and communicate over the Internet. We also use computers, personal digital assistants, cell phones, and other devices. These devices store, transmit, and retrieve information at home and at work. As we do this, we place into electronic form private, sensitive, and even incriminating information or evidence. This information is stored in various databases such as Internet-connected servers, work-related networks, and computer hard drives. This electronic trail serves as powerful legal evidence against a suspected criminal. Basically, this evidence reveals highly probative "digital fingerprints." These fingerprints can prove civil wrongs or criminal activity in a court of law.[148]

Computer Forensics. Computers and the Internet have become a pervasive element in modern life. People who engage in crime and other misconduct likewise use this technology. Effective investigation of these offenses requires evidence derived from computers, telecommunications, and the Internet. The need for digital evidence has led to a new area of criminal investigation, i.e., *computer forensics*. Forensic investigators identify, extract, preserve, and document computer and other digital evidence. Less than 20 years old, this new field is evolving rapidly. In this field, education hitherto has focused largely on technical aspects. However, investigators must deal with significant legal issues and ethical problems. Failure to follow proper legal procedure may result in the court ruling the evidence as inadmissible. As a result, a guilty criminal may go free. Failure to behave in an ethical manner erodes public confidence in law enforcement, thereby, making its job more difficult and less effective.[149]

Investigators' Pre-evidence Requirements. Forensic investigators should understand that before they seize a computer or other electronic hardware, they must consider whether the Fourth Amendment requires a search warrant. They must be aware that if they want to access stored electronic communications, they must comply with the Electronic Communication Privacy Act. If they wish to conduct real-time electronic surveillance, they need to obtain a wiretap order from a judge.[150]

---

[148] Galves, F., & Galves, C. (2006, August 25). Ensuring the admissibility of electronic forensic evidence. Galves—Professor of Law website. Retrieved April 4, 2008, from *http://www.fredgalves.com/content/view/12/28/*.

[149] Wegman, J. (2004). Computer forensics: Admissibility of evidence in criminal cases. Retrieved April 4, 2008, from *http://www.cbe.uidaho.edu/wegman/Computer%20Forensics%20AA%202004.htm*.

[150] Ibid.

Education/Training and Certification Needed. Marcus Rogers and Kate Seigfried conducted a pilot study and attempted to add to the growing body of knowledge regarding inherent issues in computer forensics. Their study consisted of an Internet-based survey that asked respondents to identify the top five issues in computer forensics. Sixty respondents answered the survey using a free-form text field. The results indicated that education/training and certification were the most reported issue (18 percent) and lack of funding was the least reported (4 percent).[151]

National Framework and Gold Standard Lacking. These findings by Rogers and Seigfried were consistent with a similar law enforcement community study by.[152] Their findings emphasize the fragmented nature of the computer forensics discipline. Currently, we lack a national framework for curricula and training development with no gold standard for professional certification. Their findings further support the criticism that there is a disproportional focus on the applied aspects of computer forensics. This disproportional focus is made at the expense of developing fundamental theories.[153],[154]

Analysis of the Admissibility of Computer Crime Evidence

In this section, I analyzed the admissibility of cyber-criminal evidence and the importance of balancing enforcement with liberty/privacy, proper handling, taking legal actions, ethical dilemmas, training, and following protocols.

Balancing Enforcement with Liberty and Privacy. Computer crime threatens our commercial and personal safety. Computer forensics has become an indispensable tool for law enforcement. However, in the

---

[151] Rogers, M. K., & Seigfried, K. (2004). The future of computer forensics: A needs analysis survey. *Computers & Security, 23(1)*, pp. 12-16.

[152] Stambaugh, H., Beaupre, D., Icove, D., Cassaday, W., & Williams, W. (2001). State and local law enforcement needs to combat electronic crime. *National Institute of Justice Research in Brief.* National Institute of Justice.

[153] Ibid.

[154] Rogers, M. K., & Seigfried, K. (2004). The future of computer forensics: A needs analysis survey. *Computers & Security, 23(1)*, pp. 12-16.

digital world, as in the physical world, the goals of law enforcement must balance with the goals of maintaining personal liberty and privacy. Computer forensic investigators must know the legal environment in which they work, or they risk the evidence they gather being ruled as *inadmissible*.[155]

Avoiding Spoliation, Tampering, and Exclusion at Trial. Evidence must be *admissible* at trial to be used to prosecute a criminal. Without a sufficient amount of credible evidence, the prosecution's case will suffer or be dismissed. Either an illegal search or seizure or a lack of authentication/foundation is an evidentiary hurdle that must be overcome to avoid suppression of evidence.[156]

Ethical Dilemmas. Computer forensic investigators face ethical dilemmas. They must exercise their discretion wisely, balancing their prosecutorial zeal with respect for citizens' individual liberties. American criminal investigators have wrestled with these same issues for over 200 years. Digital technology is not the first "new era" to challenge law enforcement. The railroad, telephone, and automobile posed similar challenges. By following the U.S. Constitutional principles and encouraging ethical behavior, we achieve the right balance between liberty and security in the digital age.[157]

Formal Training Required. Law enforcement officers, agents, investigators, prosecutors, and judges must understand technological devices. This knowledge relates to the capacity of these devices to reveal valuable and probative electronic evidence in criminal prosecutions. A way to achieve that goal is through formal training. The model used at labs like the Gulf Coast Computer Forensics Laboratory offers a good example of what we can institute at other future digital forensic labs. These labs may offer training courses

---

[155] Wegman, J. (2004). Computer forensics: Admissibility of evidence in criminal cases. Retrieved April 4, 2008, from *http://www.cbe.uidaho.edu/ wegman/Computer%20Forensics%20AA%202004.htm.*

[156] Galves, F., & Galves, C. (2006, August 25). Ensuring the admissibility of electronic forensic evidence. Galves—Professor of Law website. Retrieved April 4, 2008, from *http://www.fredgalves.com/content/view/12/28/.*

[157] Wegman, J. (2004). Computer forensics: Admissibility of evidence in criminal cases. Retrieved April 4, 2008, from *http://www.cbe.uidaho.edu/ wegman/Computer%20Forensics%20AA%202004.htm.*

that build awareness of what technological devices may contain powerful incriminating evidence. Additionally, they offer courses on how properly to search-and-seize those devices and transport them to labs for analysis. However, just as important, these labs illustrate, by example, what law enforcement officers need to know. Users electronically log in devices, and both computer and camera record all who handle them and/or transport them to different locations. Computer analysts copy the contents of these devices. Then, they extract all pertinent electronic evidence from these copies using well-developed expertise and experience.[158]

Following Protocols to Preserve Evidence. All of this potential, however, will be valuable only if prosecutors educate law enforcement investigators and technical analysts to follow the necessary protocols. In so doing, prosecutors can ensure that otherwise *admissible* electronic evidence is not suppressed or legally compromised. Inadmissibility occurs either because of an illegal search-and-seizure or because the evidentiary foundation was not properly or credibly laid during trial. Despite how criminals may employ it, technology makes life easier and more efficient for all of us. We effectively can use electronic evidence to prosecute these criminals and reduce the opportunities for criminals using technology against us. For terrorism, corporate financial scandals, computer hacking, identity theft, online fraud, or drug distribution activity, our best defense is law enforcement using criminal-activity electronic-evidence trails to bring technology-using criminals to justice.[159]

A Critique of the Cyberterrorism Problem

This section covers a critique of the cyberterrorism problem by looking at the definition of *cyberterrorism*, understanding cyber-attacks, and cyber security being a community issue.

What is Cyberterrorism? *Cyberterrorism* is the convergence of *cyberspace* and *terrorism*. It refers to unlawful attacks and threats

---

[158]   Galves, F., & Galves, C. (2006, August 25). Ensuring the admissibility of electronic forensic evidence. Galves—Professor of Law website. Retrieved April 4, 2008, from *http://www.fredgalves.com/content/view/12/28/*.

[159]   Ibid.

of attacks against computers, networks, and the information stored therein when done to intimidate or coerce a government or its people in furtherance of political or social objectives. Further, to qualify as *cyberterrorism*, an attack should result in violence against persons or property, or at least cause enough harm to generate fear. Attacks that lead to death or bodily injury, explosions, or severe economic loss would be examples. Serious attacks against critical infrastructures could be acts of *cyberterrorism*, depending on their impact. Attacks that disrupt nonessential services or that are mainly a costly nuisance would not.[160],[161]

Cyber Attacks. Terrorists may seek to cause widespread disruption and damage, including casualties, by attacking our electronic and computer networks, which are linked to other critical infrastructures such as our energy, financial, and securities networks. Terrorist groups already exploit new information technology (IT) and the Internet to plan attacks, raise funds, spread propaganda, collect information, and communicate securely. As terrorists further develop their technical capabilities and become more familiar with potential targets, cyber attacks will become an increasingly significant threat.[162] Next, we look at whether we are creating the conditions for cyber-attacks.

Terrorists Encouraged Toward Cyber-attacks? Terrorists' use of the Internet and other telecommunications devices grows both in terms of reliance for supporting organizational activities and for gaining expertise to achieve operational goals. Tighter physical and border security may also encourage terrorists and extremists to use other types of weapons to attack the United States. Persistent Internet and computer security vulnerabilities may encourage terrorists to continue to enhance their computer skills and consider a cyber attack against U.S. critical infrastructure. Alternatively, our vulnerabilities may encourage

---

[160] Denning, D. (2000, May). Testimony on "What is Cyberterrorism?" before the House Armed Services Committee.

[161] Weimann, G. (2004, December). Cyberterrorism: How real is the threat? United States Institute of Peace, Special Report No. 119. Retrieved April 4, 2008, from *http://www.usip.org/pubs/specialreports/sr119.html.*

[162] Anonymous (2002, July). *National Strategy for Homeland Security.* Washington, DC: Office of Homeland Security.

terrorists to develop alliances with criminal organizations and launch a cyber attack.[163]

A Community Issue. The cyberterrorism threat is real, and not enough people are prepared to prevent or detect an attack. Those who are unprepared affect other organizations that may be well prepared. Cyber security is a community issue. Like anything else, testing can only validate what we have already verified. However, quality professionals can be important advocates for the awareness of IT security. By keeping our eyes open and applying effective techniques, organizations will prevent attacks and recover quickly after an attack.[164]

International and Local Efforts to Combat Cyberterrorism

This section covers international and local efforts to combat cyberterrorism. We look at securing our cyber infrastructure, our national strategy to secure cyberspace, and local and international efforts to combat cyberterrorism.

Securing Our Cyber Infrastructure. To secure our cyber infrastructure against manmade and natural threats, our federal, state, and local governments work with the private sector to prevent damage, unauthorized use, and exploitation of our cyber systems. Also, we are enhancing our ability and procedures to respond in the event of an attack or major cyber incident. The *National Strategy to Secure Cyberspace* and the *National Infrastructure Protection Plan's* (NIPP's) Cross-sector Cyber Security plan guide our efforts.[165]

National Strategy to Secure Cyberspace. Following the 9/11 attacks, President George W. Bush acted quickly to secure our information and telecommunications infrastructure. The President created the Critical Infrastructure Protection Board and launched a public-private

---

[163] Rollins, J., & Wilson, C. (2007, January 22). Terrorist capabilities for cyberattack: Overview and policy issues. *CRS* [Congressional Research Service] *Report for Congress*, Order Code RL33123, 28 pp.

[164] Rice, R. W. (n.d.). The threat of cyberterrorism. Rice Consulting Services, Inc. Retrieved April 4, 2008, from *http://www.riceconsulting.com/articles/threat-of-cyberterrorism-pf.html*.

[165] Anonymous (2007, October). *National Strategy for Homeland Security*. Washington, DC: Homeland Security Council.

partnership to create a *National Strategy to Secure Cyberspace.* Issued in February 2003, this National Strategy provides a roadmap to empower all Americans to secure the part of cyberspace they control, including a variety of new proposals aimed at the following five levels: (1) the home user and small business, (2) large enterprises, (3) sectors of the economy, (4) national issues, and (5) global issues.[166]

Local Efforts. Thousands of citizens all across the country have contributed to the effort by contributing their views in Town Hall meetings, on interactive websites, or by participating in one of the dozens of participating groups and associations. State and local governments and state and local law enforcement have also united to prepare their own cyber security strategies.[167] Recently, I certified and became a member of the San Marcos (California) Community Emergency Response Team (CERT) as well as a disaster services worker (DSW) with the Unified San Diego County Emergency Services Organization (USDCESO). I am a counterterrorism specialist with the San Marcos CERT doing my little part to combat cyberterrorism at the local community level.

International Efforts. Cybercrime is a major international challenge; however, attitudes about what comprises a criminal act of computer wrongdoing may still vary from country to country. The European Union has established the Critical Information Infrastructure Research Coordination Office (CI2RCO), which is responsible to examine how its member states are protecting their critical infrastructures from possible cyber attacks. The project identifies research groups and programs focused on IT security in critical infrastructures.[168]

A consultative assembly of 43 countries, the Council of Europe, based in Strasbourg, France, adopted the Convention on Cybercrime in 2001. The Convention, which became effective in July 2004, is the *first and only* international treaty to deal with breaches of law "over the Internet or other information networks." The Convention requires

---

[166] Anonymous (2002, July). *National Strategy for Homeland Security.* Washington, DC: Office of Homeland Security.

[167] Ibid.

[168] Rollins, J., & Wilson, C. (2007, January 22). Terrorist capabilities for cyberattack: Overview and policy issues. *CRS* [Congressional Research Service] *Report for Congress,* Order Code RL33123, 28 pp.

participating countries to update and harmonize their criminal laws against hacking, infringements on copyrights, computer-facilitated fraud, child pornography, and other illicit cyber activities.[169],[170] To date, 8 of the 42 countries that signed the Convention have already completed the ratification process.[171]

An Evaluation of the Effectiveness of U.S. Infrastructure Programs

In this section, I evaluated the effectiveness of U.S. infrastructure programs by looking at the sectors of critical infrastructure and key resources (CI/KR), pervasiveness of IT, fear, and collective security approach and model.

Sectors of CI and KR. Our Nation has identified 17 sectors of CI and KR, each with cross-cutting physical, cyber, and human elements. These 17 sectors include (1) Agriculture and Food, (2) Banking and Finance, (3) Chemical, (4) Commercial Facilities, (5) Commercial Nuclear Reactors, Materials, and Waste, (6) Dams, (7) Defense Industrial Base, (8) Drinking Water and Water Treatment Systems, (9) Emergency Services, (10) Energy, (11) Government Facilities, (12) Information Technology, (13) National Monuments and Icons, (14) Postal and Shipping, (15) Public Health and Health Care, (16) Telecommunications, and (17) Transportation Systems.[172]

IT is Interwoven Through all CIs. Information technology is essential to virtually all of the nation's CIs from the air traffic control system to the aircraft themselves, from the electric power grid to the financial and banking systems, and, obviously, from the Internet to communication systems. Hence, this reliance of all of the nation's CIs

---

[169] Anonymous (2001, November 23). Convention on cybercrime. Council of Europe, ETS [European Treaty Series] No. 185, Budapest, 25 pp.

[170] Rollins, J., & Wilson, C. (2007, January 22). Terrorist capabilities for cyberattack: Overview and policy issues. *CRS* [Congressional Research Service] *Report for Congress*, Order Code RL33123, 28 pp.

[171] Ibid.

[172] Anonymous (2007, October). *National Strategy for Homeland Security*. Washington, DC: Homeland Security Council.

on IT makes all of them vulnerable to sabotage through their computer or telecommunication systems.[173],[174]

Fear of Cyberterrorism. Policymakers and the public today see the threat of a terrorist attack on IT, or "cyberterrorism," as being one of the greatest dangers to the United States. As the first director of the Department of Homeland Security, Tom Ridge,[175],[176] warned in April 2002, a "terrorist can sit at one computer connected to one network and can create worldwide havoc." The notion of cyberterrorism links the fear of random, violent victimization with the distrust and outright fear of computer technology.[177],[178] This union compounds the fear of cyberterrorism.[179]

Collective Security Approach Needed. Protecting the CI from cyber attack is of great import to nations today. It is also important for global security and prosperity. In recent years, the threat posed has changed from what once appeared as an unstructured threat from adventurous hackers, to a structured, hostile attack on elements of the CIs of different countries. In some cases, governments and organizations with substantial

---

[173] National Research Council (2003). *Information Technology for Counterterrorism: Immediate Actions and Future Possibilities.* Washington, DC: National Academies Press, p. 2.

[174] Gershman, J. (2004, September). A Secure America in a Secure World. *Foreign Policy in Focus (FPIF) Special Report.* Silver City, New Mexico: Interhemispheric Resource Center (IRC), 29 pp.

[175] Ridge, T. (2002, April 23). Remarks by Homeland Security Director Tom Ridge to the Electronics Industries Alliance. Retrieved at *http://www.eia. org/events/springconf/remarks_ridge_1.phtml.*

[176] Gershman, J. (2004, September). A Secure America in a Secure World. *Foreign Policy in Focus (FPIF) Special Report.* Silver City, New Mexico: Interhemispheric Resource Center (IRC), 29 pp.

[177] Pollitt, M. (1997, October). Cyberterrorism—fact or fancy. Proceedings of the 20th National Information Systems Security Conference. Retrieved April 5, 2008, from *http://www.cs.georgetown.edu/~denning/infosec/ pollitt.html.*

[178] Gershman, J. (2004, September). A Secure America in a Secure World. *Foreign Policy in Focus (FPIF) Special Report.* Silver City, New Mexico: Interhemispheric Resource Center (IRC), 29 pp.

[179] Ibid.

resources are increasingly backing such attacks. To respond properly to this threat to security and prosperity, we need a strong, international solution grounded in a political framework, for isolated technical or legal solutions will not work. Moreover, efforts to confront structured hostile threats on a national level have been less than successful, and the technology employed has not been adequate to seal the systemic vulnerabilities in the IT-dependent CI. We need a collective security approach to protect the global CI.[180]

International Treaty Approach Goes Wanting. Dr. Stephen Bryen, managing director of Aurora Defense, concluded that much important security work has been carried out internationally in a wide variety of forums. However, efforts have fallen short of a formal international treaty approach to the problem. The key benefit of such an approach is that it represents a political means to combat a problem, which threatens the CI of many countries. Furthermore, ultimately, it may undermine collective global security and peace.[181]

Collective Security Model. The collective security model—in the form of a treaty organization—is an appropriate response to a structured hostile threat from cyberterrorists. In today's environment of increasing connectivity of global network systems, the framework described above bears the essential features that such an organization would require if we expect to challenge successfully the problem of structured cyberterrorism. For nations to protect adequately their future critical network infrastructures, a need exists to consider such a collective approach as the way forward.[182]

Conclusion

Technology makes life easier and more efficient for all of us. However, computer crime threatens our commercial and personal

---

[180] Bryen, S. (2002, May 20). A collective security approach to protecting the global critical infrastructure. The author discussed this paper at the ITU [International Telecommunication Union] Workshop on Creating Trust in Critical Network Infrastructures, held in Seoul, Republic of Korea, on 20-22 May 2002, Document: CNI/09, 17 pp.

[181] Ibid.

[182] Ibid.

safety. Computer forensics has become an indispensable tool for law enforcement. In the digital world, as in the physical world, the goals of law enforcement must balance with the goals of maintaining personal liberty and privacy. American criminal investigators have wrestled with these same issues for over 200 years. Currently, we lack a national framework for curricula and training development with no gold standard for professional certification.

Cybercrime is a major international challenge. We need a collective security approach to protect the global CI. The Convention on Cybercrime, which became effective in July 2004, is the *first and only* international treaty to deal with breaches of law over the Internet or other information networks. Policymakers and the public today see the threat of a terrorist attack on IT, or *cyberterrorism*, as being one of the greatest dangers to the United States. Cyber security is a community issue. By keeping our eyes open and applying effective techniques, organizations will prevent attacks and recover quickly after an attack. However, we still are not where we must be to possess effective infrastructure programs in the United States.

Chapter 6
Critical Infrastructure Protection Program

Robert T. Uda, MBA, MS

In this chapter, I identify a critical infrastructure protection (CIP) program that is being implemented at the local government level, and I have evaluated what weaknesses the program may possess. I selected the City of North Miami Beach, Florida, for this analysis.

Local governments represent the front lines of protection and the face of public services to the American people. Their core competencies must include knowledge of their communities, residents, landscapes, and existing critical services for maintaining public health, safety, and order. Communities look to local leadership to assure safety, economic opportunities, and quality of life (QOL).

Public confidence, therefore, starts locally and is dependent upon how well communities plan and protect their citizens, respond to emergencies, and establish order out of chaos. To this end, the City of North Miami Beach Police Department, the private sector, and concerned citizens have begun an important partnership and commitment to action.

To introduce you to The City of North Miami Beach, Florida, CIP program being implemented and to evaluate the weaknesses of that program, this chapter covers the following topics:

- What is a Critical Infrastructure?
- What is the Critical Infrastructure Protection Program (CIPP)?
- Examples of Local Government CIPPs
- Selecting a CIPP Implemented at the Local Government Level
- Evaluating the Selected CIPP Implemented at the Local Government Level

- Weaknesses of the Selected CIPP
- What Needs to Be Done to Improve on the Identified Weaknesses

What is a Critical Infrastructure?

Definition. Our nation's economic vitality, national security, and QOL of our citizens depend upon the availability, continuous operation, and reliability of several different infrastructure sectors . . . both physical and "virtual." Since these various sectors provide a framework around which we live our daily lives, conduct business, and function as a society, we deem these sectors as critical to our country's existence. The attacks of September 11, 2001, heightened awareness of our nation's vulnerabilities. We realize the necessity to secure our critical infrastructure (CI) from future terrorist attacks. Additionally, we need to secure our CI from all disasters and events that could disrupt and threaten our way of life.[183]

Examples of Unique Protection Areas. The following individual CI sectors and special categories of key resources (KRs) have unique issues that require protective action.[184]

(a) *Examples of CIs:* The following CI sectors need protection: (1) Agriculture and Food, (2) Drinking Water and Water Treatment Systems, (3) Public Health and Healthcare, (4) Emergency Services, (5) Defense Industrial Base, (6) Information Technology (IT), (7) Telecommunications, (8) Energy, (9) Transportation Systems, (10) Banking and Finance, (11) Chemical, and (12) Postal and Shipping.

(b) *Examples of KRs:* The following KR categories also need protection: (1) National Monuments and Icons; (2) Commercial Nuclear Reactors, Materials, and Waste; (3) Dams; (4) Government Facilities; and (5) Commercial Facilities.

---

[183] Hopkins, B. (2003). *State Official's Guide to Critical Infrastructure Protection.* Lexington, Kentucky: The Council of State Governments, 100 pp.

[184] Anonymous (2007, October). *National Strategy for Homeland Security.* Washington, DC: Homeland Security Council.

What is the Critical Infrastructure Protection Program (CIPP)?

Implementing Protective Programs. The risk assessment and prioritization process enable the Department of Homeland Security (DHS), sector-specific agencies (SSAs), and other security partners to identify opportunities. These opportunities enhance current CI/KR protection programs where they will offer the greatest benefit. Security partners give priority to developing CI/KR protection programs that focus resources on assets, systems, networks, and functions considered to be at greatest risk.[185]

Local Governments. Local governments represent the front lines for homeland security and, more specifically, for CI/KR protection and implementation of the National Infrastructure Protection Plan (NIPP) partnership model. They provide critical public services and functions in conjunction with private sector owners and operators. In some sectors, local government entities own and operate CI/KR such as water, storm-water, and electric utilities. Malevolent acts or disruptions that impact CI/KR begin and end as local situations.[186]

Local Authorities and Security Partners. Local authorities usually handle initial prevention, response, and recovery operations until coordinated support from other sources becomes available. They take this operational approach regardless of who owns or operates the affected asset, system, or network. As a result, local governments are critical partners under the NIPP framework. They drive emergency preparedness and local participation in implementing NIPP and sector-specific plans (SSPs) across a variety of jurisdictional security partners. These security partners include government agencies, owners/operators, and private citizens in the served communities.[187]

Coordinated Effort. State, local, and tribal authorities are responsible for providing or augmenting protective actions for assets, systems, and networks that are critical to the public within their jurisdiction and authority. They develop protective programs,

---

[185] Anonymous (2006a). *National Infrastructure Protection Plan*. Washington, DC: Department of Homeland Security, 196 pp.

[186] Ibid.

[187] Ibid.

supplement federal guidance and expertise, implement relevant federal programs, and provide specifically needed law enforcement capability. When appropriate, they access federal resources to meet jurisdictional protection priorities.[188]

Examples of Local Government CIPPs

City of North Miami Beach, Florida, CIPP. Through its CIPP, the North Miami Beach Police Department continues to address the threat of terrorism and any potential attacks against CIs within the community. By strengthening prevention and preparedness efforts against terrorist attacks, major disasters, and other emergencies, the city government develops appropriate levels of response to these types of incidents. The Police Department continually assesses the vulnerabilities of the CIs and increases investment in their security. They do this by continuously taking required steps to identify and protect KRs and services within their jurisdiction.[189]

The success of protective efforts requires close cooperation among government, community, and private sector participants at all levels. Each citizen has an important role to play in protecting the CIs/KRs, which are basic to their daily lives. Strong partnerships with citizens groups and organizations provide support to prevent terrorist incidents. Groups such as Neighborhood Watch and Police Citizens Patrol enhance prevention efforts by being an extra set of eyes and ears. They identify and report strange or suspicious behavior as related to terrorist activity. The North Miami Beach Police Department was one of the first agencies in the country to offer local citizens with specialized training in Community Terrorism Preparedness and Strategic Actions for Emergency Response or SAFER.[190]

---

[188] Ibid.

[189] Anonymous (n.d.). City of North Miami Beach, Florida, Critical Infrastructure Protection Program. The official website of the City of North Miami Beach, Florida. Retrieved April 18, 2008, from *http://northmiamibeach. govoffice.com/index.asp?Type=B_BASIC&SEC=%7BF04CACB1-1126-4FF5-85C3-81ACF84754F9%7D.*

[190] Ibid.

George Mason University CIPP. The George Mason University CIPP (of Fairfax, Virginia) studied the effects of the 2004 and 2005 tropical storm seasons. They were particularly interested in the infrastructure damage caused by hurricanes in Florida, Louisiana, Mississippi, and Texas. The CIPP study recommended that federal and state agencies find ways to better share information before, during, and after a catastrophe. Large disasters, such as the 2005 mega-storms, do not respect political borders. Thus, it is up to state and local governments and the federal government to plan better and to coordinate responses well before, during, and long after the initial emergency.[191]

Skilled utility workers, technicians, and power engineers comprise the nation's critical human infrastructure, which is an aging and shrinking collective workforce. Faculty and teachers at post-secondary educational institutions experience similar attrition. Additionally, research money is diminishing. The twin challenges of education inputs/outputs and the aging of a knowledgeable workforce present an emerging issue directly related to the ability to increase the grid's reliability and respond to future large-scale catastrophes.[192]

Kentucky CIPP. Through the Kentucky Homeland Security University Consortium, the National Institute for Hometown Security (NIHS) manages the Kentucky Critical Infrastructure (KCI) Protection Program. The program focuses on research, development, and technology transfer designed to protect the nation's community-based CIs. In consultation with the U.S. Department of Homeland Security (DHS), NIHS determines research needs and requirements. The DHS then provides program funding to perform those research needs and requirements.[193]

In previous years, the KCI provided funding for 23 separate research and development (R&D) projects. Each project requires a

---

[191] Driver, D. (2007, January 16). Critical infrastructure protection program examines effects of 2005 hurricanes. *The Mason Gazette*. Retrieved April 18, 2008, from *http://gazette.gmu.edu/articles/9559/print.*

[192] Ibid.

[193] Anonymous (2007, August). Kentucky critical infrastructure protection program: Request for research and development project proposals. The National Institute for Hometown Security. Retrieved April 18, 2008, from *http://www.thenihs.org/program_kci_criteria.php.*

Kentucky-based university serving as the lead institution for the research, development, and coordination activities. DHS expects at least one other Kentucky-based college or university as well as non-Kentucky universities and institutions to participate on each project. Where appropriate, DHS also encourages private sector participation.[194]

Because of the successful KCI experience with the existing 23 projects, DHS has decided to fund and support a more advanced structure. This advanced structure allows performance of more relevant and integrated technology solutions. The approach moves the program towards national significance and long-term stability. Participating universities and principal investigators (PIs) on previous KCI projects will notice significant differences in the current solicitation and subsequent project management process.[195]

National Capital Region. Following the events of September 11, 2001, protecting CIs has become a major concern in the United States. Infrastructure systems are critical because they are essential to the functioning of modern societies. Additionally, their network characteristics make them vulnerable to both random disasters and intentional attacks. Planning for and response to large-scale disruptions require coordination among local, state, and federal authorities and also the private sector, which are the primary owners of such infrastructures.[196]

This interdisciplinary research project presents published findings that evaluate CIP measures and their impact on security-related cooperation within the National Capital Region (Washington, DC metro area). Prior research on governance options for regional infrastructure systems and on inter-agency disaster response capabilities indicates a variety of possible modes of coordination.[197]

---

[194]   Ibid.

[195]   Ibid.

[196]   Pommerening, C. (2005, November 3-5). Governance structures for protecting critical infrastructures: The example of the National Capital Region. Paper presented at the 2005 Fall Research Conference of the Association for Public Policy Analysis and Management (APPAM), Washington, DC. Retrieved April 18, 2008, from *http://cipp.gmu.edu/clib/pubDetail.php?id=84.*

[197]   Ibid.

CIP for the Local Naval Installation. Hun Kim wrote that the Presidential Decision Directive-63 (PDD-63) recognizes the growing potential vulnerability of physical and cyber-based systems essential to the economy and government. As a result, the Department of the Navy (DON) designed the CIP program to develop, administer, and coordinate an enterprise-wide effort to do the following:

- Identify mission essential infrastructures
- Assess their vulnerability
- Develop a coordinated physical and cyber indications and warning capability
- Take the necessary actions to ensure achievement of DON objectives during CI loss

Begun in 1999, this DON CIP program actively pursues outreach to regional and local commanders for education on the CIP construct to coordinate Naval Integrated Vulnerability Assessments.[198]

Selecting a CIPP Implemented at the Local Government Level

Selection Process. Since computer security, forensics, and cyberterrorism are of prime interest, it is only natural to select the CIPP that covers the IT and Communications sectors. Moteff, Copeland, and Fischer define this sector as computing and telecommunications equipment, software, processes, and people that support the:[199]

- Processing, storage, and transmission of data and information
- Processes and people that convert data into information and information into knowledge

---

[198] Kim, H. (n.d.). Critical infrastructure protection for the local installation. Retrieved April 18, 2008, from *http://www.chips.navy.mil/archives/02_Summer/authors/index2_files/cip.htm.*

[199] Moteff, J., Copeland, C., & Fischer, J. (2003, January 29). Critical infrastructures: What makes an infrastructure critical? Report for Congress, Congressional Research Service, The Library of Congress, Order Code RL31556, 20 pp.

- Data and information themselves

The Selected CIPP. I selected the City of North Miami Beach, Florida, CIPP. Of all considered local government programs, the City of North Miami Beach appears to possess the most comprehensive CIPP of all. They have an excellent program. However, there still is room for improvement.

Evaluating the Selected CIPP Implemented at the Local Government Level

Securing the Community. There is an old saying in business that directly applies to our concerns about critical infrastructure assurance—"think globally, act locally." In the final analysis, all disasters are local. That was certainly the case on September 11, 2001. If we are truly to secure the nation's homeland, we must proceed by securing our respective communities, one by one.[200]

Indeed, the community represents an essential focal point for building a foundation for any national initiative that seeks to influence the public. All of us live in communities and look to the community for leadership to assure our safety, our economic opportunities, and our QOL. The first people on the scene in a crisis are not the feds . . . but the local emergency response teams. We need leadership at the local level to survive terrorist attacks. Public confidence starts in the community and is dependent on how well the community plans ahead, responds to crises, and reestablishes order from chaos. It was not by accident that the public "face" in New York and Washington, DC, on September 11 and in the days thereafter, was the mayor of each of these cities.[201]

A Different Viewpoint. It is interesting to note that Lewis & Darken take issue with Juster and the "think globally, act locally" strategy. They

---

[200] Juster, K. I. (2002, February 13). Homeland security and critical infrastructure assurance: The importance of community action. Remarks of the Undersecretary of Commerce for Export Administration at the conference on Critical Infrastructures: Working Together in a New World, Austin, Texas. Retrieved April 18, 2008, from *http://www.bis.doc.gov/ news/2002/communityactionimportantnhomelandsecurity.htm.*

[201] Ibid.

say that this policy is not only dangerous—because local jurisdictions will never have the capability to protect their CI assets—but an unfortunate waste of money.[202] In fact, Paul Posner of the Government Accountability Office (GAO) recognized this problem soon after the formation of the DHS and testified to the U.S. Senate Committee on the Judiciary, "The challenges posed in strengthening homeland security exceed the capacity and authority of any one level of government."[203]

Re-evaluation of National Strategy Needed. It is time to re-evaluate the national strategy and replace state and local strategies with a national effort. The Department of Interior and Forest Service have done this: A federal force has mainly fought large forest fires across regional boundaries. The food and agriculture sector has done this to some extent. For example, Food and Drug Administration (FDA) regulators work with the private sector to ensure the safety of the food supply. Additionally, whether or not we admit it, the Federal Bureau of Investigation (FBI) is a national police force that transcends state and local borders.[204] Lewis & Darken make a reasonable case for rethinking of our strategy from one where the current local government runs the show with support from the federal government to a strategy where the federal government runs the show with support from local governments. It seems feasible and reasonable for broad, wide-area disasters. For strictly local disasters, however, the local government would best deal with them.

Weaknesses of the Selected CIPP

Telecommunications/IT Sector Challenges. All levels of government are working together to address vulnerabilities of the telecommunications and IT sectors. However, state and local governments

---

[202] Lewis, T. G., & Darken, R. (2005). Potholes and detours in the road to critical infrastructure protection policy. *Homeland Security Affairs, 1(2)*.

[203] Posner, P. (2003, September 3). Homeland security: Reforming federal grants to better meet outstanding needs. Testimony before the Subcommittee on Terrorism, Technology, and Homeland Security; Committee on the Judiciary, US Senate. GAO-03-1146T.

[204] Lewis, T. G., & Darken, R. (2005). Potholes and detours in the road to critical infrastructure protection policy. *Homeland Security Affairs, 1(2)*.

face special challenges related to working with the private sector. These special challenges include addressing vulnerabilities in the nation's computer-controlled systems and developing mechanisms and processes to protect them from attack. The private sector plays a central role in securing cyberspace because it depends on this infrastructure to conduct business. Additionally, it owns and operates the vast majority of the infrastructures and cyber systems upon which the nation depends.[205]

Interdependency of Communications and IT. Many organizations rely on computers and the Internet for day-to-day operations (both on-site and remote), delivery of services, data management, and marketing. This situation illustrates the difficulty in separating physical CIP from cyber CIP and separating the communications sector from all the other sectors. As a result, safeguarding communications and IT is essential to assuring the survival of CIs. Home users and small businesses can contribute to cyber security by using safe passwords, maintaining and updating virus protection and patches, and using traffic filtering, firewalls, and similar good practices.[206]

Not Enough Effort Being Expended. In 2002, Richard Clarke, chairman of the president's Critical Infrastructure Protection Board, said, "I'm never satisfied. I'm feeling good about the federal government's own activities and that major sectors of the private sector are taking action. For example, the banking and finance sector is doing a great deal; the electric power grid is for the first time thinking about encryption; and the IT sector itself is beginning to talk about quality software development and making security a design criteria. Companies like Oracle [Corporation], Sun [Microsystems Inc.], Microsoft [Corporation], and Cisco [Systems Inc.] are leading that effort. IT security is also a top issue in the private sector."[207]

---

[205] Hopkins, B. (2003). *State Official's Guide to Critical Infrastructure Protection*. Lexington, Kentucky: The Council of State Governments, 100 pp.

[206] Anonymous (2008, February 12). Best practices for the protection of critical infrastructure. Public Safety and Emergency Preparedness Canada. Retrieved April 18, 2008, from *http://www.publicsafety.gc.ca/prg/em/nciap/best_practices-en.asp*.

[207] Verton, D. (2002, September 6). White House cyber security chief defines cyberthreat. *Computerworld*. Retrieved on April 18, 2008, from *http://*

What Needs to Be Done to Improve on the Identified Weaknesses

Fund the Private Sector. Since the private sector owns and operates most of the nation's computer and cyber systems, it is only natural that the private sector takes the lead role in protecting their own systems. The state and local governments follow the lead of the private sector. State and local governments can contribute to a greater extent in protecting the IT and telecommunications CI sectors. They can do this by contributing funds to help the private sector do a better job in defending their systems from those who intend to harm them.

An Integrated System. The telecommunications and IT sectors are inseparably interwoven with most of the other CI sectors but more so with each other. Rather than separating the physical CIP from the cyber CIP, we should take the "systems" approach by dealing with them as an integrated system. If anything, they should be dealt with as a "system of systems."

Additional Initiatives. Since 2002, much has been done to overcome the weaknesses of a CIPP at the local government level. However, even more can be done. Additional initiatives that can/should be taken include the following:[208]

- Strengthen interoperable communications capabilities
- Enhance planning infrastructure capabilities to ensure preparedness for terrorism and all hazard events
- Strengthen information sharing and collaboration capabilities
- Establish a CIP for the state

Conclusion

Terrorists seek to undermine confidence in our public and private institutions and in our ability to manage the consequences of their

---

www.computerworld.com/action/article.do?command=printArticleBasic &articleId=74033.

[208] Anonymous (2006). State enhancement plan. Retrieved April 24, 2008, from *http://www.oregon.gov/OPS/CJS/docs/Homeland_Security/2006_ State_Enhancement_Plan.pdf.*

attacks. In response, the federal government must work collaboratively and in partnership with state and local governments, with the private sector, and with local citizens. To the extent that government and private industry are believed to be doing everything within reason to protect the public from harm, the public's confidence in its institutions will remain intact despite such attacks. We can have the best national strategy for homeland security that the most brilliant minds in Washington can devise, and yet we will fail in our endeavor if local communities do not meet the immediate challenges of a terrorist disaster.[209]

Local governments represent the front lines of protection and the face of public services to the American people. Their core competencies must include knowledge of their communities, residents, landscapes, and existing critical services for maintaining public health, safety, and order. Communities look to local leadership to assure safety, economic opportunities, and QOL. Public confidence, therefore, starts locally and is dependent upon how well communities plan and protect their citizens, respond to emergencies, and establish order out of chaos.[210] To this end, the City of North Miami Beach Police Department, the private sector, and concerned citizens have begun an important partnership and commitment to action.[211]

---

[209] Juster, K. I. (2002, February 13). Homeland security and critical infrastructure assurance: The importance of community action. Remarks of the Undersecretary of Commerce for Export Administration at the conference on Critical Infrastructures: Working Together in a New World, Austin, Texas. Retrieved April 18, 2008, from *http://www.bis.doc.gov/ news/2002/communityactionimportantnhomelandsecurity.htm*.

[210] Anonymous (2003, February). *The National Strategy for the Physical Protection of Critical Infrastructures and Key Assets*. Washington, DC: The White House.

[211] Anonymous(n.d.).CityofNorthMiamiBeach,Florida,CriticalInfrastructure Protection Program. The official website of the City of North Miami Beach, Florida. Retrieved April 18, 2008, from *http://northmiamibeach. govoffice.com/index.asp?Type=B_BASIC&SEC=%7BF04CACB1-1126-4FF5-85C3-81ACF84754F9%7D*.

Chapter 7

Recommendations on Cyber Threats and Warfare

Robert T. Uda, MBA, MS

T his chapter covers what I consider to be the five most important recommendations in dealing with cyber threats and cyberwarfare. These five most important recommendations include the following:

1. ***Defense Also Involves Offense.*** There needs to be an understanding that defense also involves offense.
2. ***Tools for Offensive Online Warfare.*** It is unclear whether the tools for offensive online warfare exist, and extensive R&D efforts may be needed to improve them.
3. ***Central Organization in the DoD to Fight Cyberwarfare.*** There must be a clear central organization in the DoD that is trained and equipped to fight cyberwarfare and respond to large-scale foreign attacks by governments, terrorists, and extremists.
4. ***A Clear Response Doctrine.*** The U.S. government needs to develop a clear response doctrine.
5. ***Right to Respond Unilaterally.*** The United States should reserve the right to respond unilaterally to attacks against its infrastructure.

The ensuing sections detail each of my five most important recommendations for homeland defense. The reason why I consider these five as the most important recommendations is because they are all tightly interrelated and work together as an integrated entity.

Defense Also Involves Offense

Recommendation No. 1. *There needs to be an understanding that defense also involves offense.*

A *cyber war* is a conflict that uses hostile, illegal transactions or attacks on computers and networks in an effort to disrupt communications and other pieces of infrastructure as a mechanism to inflict economic harm or upset defenses.[212] This is one of the most important recommendations because if our enemies conduct a cyber war against us, we must be prepared to defend ourselves and also be capable of taking the offensive. *The best defense is a good offense.*

Cyber War Games. To prepare for a cyber war, the U.S. government conducted a series of cyber war games to test its ability to recover from and respond to digital attacks. Code-named 'Cyber Storm II,' this was the largest-ever exercise that evaluated the mettle of information technology (IT) experts and incident response teams from 18 federal agencies, which included the Central Intelligence Agency (CIA), Department of Defense (DoD), Federal Bureau of Investigation (FBI), and National Security Agency (NSA). Additionally, it included officials from nine states including Delaware, Pennsylvania, and Virginia. Furthermore, over 40 companies participated including Cisco Systems, Dow Chemical, McAfee, and Microsoft.[213]

In the inaugural Cyber Storm two years ago, planners simulated attacks against the communications and IT sector and the energy and airline industries. This year's exercise featured mock attacks by nation states, terrorists, and saboteurs against the IT and communications sector and the chemical, pipeline, and rail transportation industries.[214]

Jerry Dixon, a former director of the National Cyber Security Division at the Department of Homeland Security (DHS) who helped

---

[212] Coleman, K. (2008, January 28). Coleman: The cyber arms race has begun. *Csoonline.com.* Retrieved May 3, 2008, from *http://www.csoonline.com/article/print/216991.*

[213] Krebs, B. (2008, March 19). White House taps tech entrepreneur for cyber defense post. *Washingtonpost.com.* Retrieved May 4, 2008, from *http://homelandsecurity.osu.edu/focusareas/cyberterrorism.html.*

[214] Ibid.

to plan both exercises, said Cyber Storm is designed to be a situational pressure-cooker for players. Those who adopt the proper stance or response to a given incident are quickly rewarded by needing to respond to even more complex and potentially disastrous scenarios.[215]

Offensive Cyber War. Lt. Gen. Robert Elder Jr., USAF, has revealed that a U.S. Air Force Cyber Command is set to become operational in October 2008. It is aimed at not only fighting off "cyber" attacks from foreign countries and terrorist groups but also to go on the offensive. According to General Elder, offensive cyber-attacks in network warfare make conventional attacks more effective, e.g., if an adversary's integrated defense systems or weapon systems can be disrupted using cyber attacks. Modern armies that tend to be more and more dependent on computers and computer networks also become more vulnerable to network attacks.[216]

U.S. Cyberwarfare on the Offensive. The United States, of course, is no innocent bystander. William M. Arkin, a defense analyst who writes the *Early Warning* blog for the *Washington Post*, says, "Our ability to penetrate into enemy computer networks, our ability to exploit communication networks, to manipulate digital information, is real," but little is known about the precise nature of Washington's offensive capabilities. Some details, however, have leaked. For instance, in March 2004, the Pentagon announced the formation of an Information Operations team—the Network Attack Support Staff—to streamline the military's cyber attack capabilities. The aim, senior military officials said at the time, was to create an "interface between the combatant commanders and the intelligence community."[217]

Offensive Information Warfare. The DoD maintains redundant systems in place to defend its network against cyber attacks. However, in 2007, it has started to push development of offensive information warfare capabilities. If "we apply the principle of warfare to the cyber

---

[215] Ibid.

[216] Anonymous (2008, April 13). Cyber war. *The Sunday Times Online*, 42(46). Retrieved May 3, 2008, from *http://www.sundaytimes.lk/080413/ Mirror/mirrorTechnoPage.html*.

[217] Bruno, G. (2008, February 27). The evolution of cyberwarfare. Council on Foreign Relations. Retrieved May 3, 2008, from *http://www.cfr.org/ publication/15577/evolution_of_cyber_warfare.html*.

domain, as we do to sea, air, and land, we realize the defense of the nation is better served by capabilities enabling us to take the fight to our adversaries, when necessary, to deter actions detrimental to our interests," Marine Gen. James Cartwright, commander of the Strategic Command, told the House Armed Services Committee in March 2007.[218]

Tools for Offensive Online Warfare

Recommendation No. 2. *It is unclear whether the tools for offensive online warfare exist, and extensive R&D efforts may be needed to improve them.*

This recommendation is most important because if we are to take the offensive, we must possess offensive tools. Additionally, if we do not possess sufficient offensive tools, then we must conduct the R&D to develop and improve them.

Cyberwarfare Strategies. Technology experts and military strategists as well as city and urban planners are collaborating on cyberwarfare strategies design to disrupt and defend against critical offensive and defensive operations. The Naval Postgraduate School has defined the following three levels of offensive cyber capabilities:[219]

1. *Simple-Unstructured:* the capability to conduct basic hacks against individual systems using tools created by someone else.
2. *Advanced-Structured:* the capability to conduct more sophisticated attacks against multiple systems and possibly to modify or create basic tools.
3. *Complex-Coordinated:* the capability for coordinated attacks capable of causing mass-disruption against many defense systems.

---

[218] Brewin, B. (2007, October 24). Management matters: Cyber wars. *GovernmentExecutive.com*. Retrieved May 3, 2008, from *http://www.gov exec.com/story_page_pf.cfm?articleid=38352&printer friendlyvers=1.*

[219] Coleman, K. (2008, January 28). Coleman: The cyber arms race has begun. *Csoonline.com*. Retrieved May 3, 2008, from *http://www.csoonline.com/article/print/216991.*

Using a combination of the above levels of capabilities, cyber war plans are emerging and driving the need for a wide range of cyber weapons.[220]

Offensive Technologies. William M. Arkin, a defense analyst who has reported on cyber security issues for over two decades, says the U.S. military also has technologies capable of penetrating and jamming enemy networks including the classified "Suter" system of airborne technology. According to *Aviation Week*, the military has integrated Suter into unmanned aircraft and "allows users to invade communications networks, see what enemy sensors see, and even take over as systems administrator so sensors can be manipulated into positions so that approaching aircraft can't be seen." Some speculate the Israeli military used the capability during its air raid on a Syrian construction site in September 2007. The United States made use of nascent capabilities in the 1999 Kosovo War and built on those lessons in Iraq.[221]

Cyber Weapons. In the mid-1990s, a study by the RAND Corporation illustrated the costs of developing the cyber weapons needed for conducting cyber war are extremely modest. That being the case, almost every country can afford these efforts. Lani Kass, a senior adviser to USAF Chief of Staff Gen. T. Michael Moseley, emphasized the need for the United States to develop an offensive cyber capacity. Cyber arms are seen as first-strike weapons used to disrupt the enemy's command, control, and operational infrastructure and possibly create civil unrest through interrupting basic infrastructure and services. In a report developed by Spy-Ops in the fall of 2007, they estimated that about 140 countries possess in place and operational active cyber weapons development programs.[222]

Examples of Offensive Cyber Weapons. Some of the most common types of offensive cyber weapons include the following: (1) wireless

---

[220] Ibid.

[221] Bruno, G. (2008, February 27). The evolution of cyberwarfare. Council on Foreign Relations. Retrieved May 3, 2008, from *http://www.cfr.org/ publication/15577/evolution_of_cyber_warfare.html.*

[222] Coleman, K. (2008, January 28). Coleman: The cyber arms race has begun. *Csoonline.com*. Retrieved May 3, 2008, from *http://www.csoonline.com/ article/print/216991.*

network disruptors, (2) computer viruses, (3) transient electromagnetic devices, (4) malware, (5) transaction generators, (6) contaminated software, (7) Trojan horse software applications, (8) hacker kits/root kits, (9) worms, (10) key loggers, (11) spyware, (12) password crackers, (13) encryption crackers, (14) logic bombs, (15) back doors, (16) spoofing. We should note that exploitation kits have been developed for cell phones and game stations to use these devices in distributed denial of service attacks and to steal data.[223]

Cyber War. Offensive cyber weapons have been developed by multiple countries that could create havoc and damage to our information infrastructure. Cyber arms have become easier to obtain, easier to use, and much more powerful. These weapons are a fraction of the cost of traditional weapons such as a tank. Therefore, state—or group-sponsored attacks against our information systems using computer viruses and other techniques should be considered an act of war. As such, governments must be proactive and establish parameters, definitions, and regulations around cyber war.[224]

Research and Development. The Army and Air Force started pushing to acquire technology to go on the offense in cyberspace in 2007. In May 2007, Army officials released a solicitation for a wide range of offensive information tools, saying, "technologies designed to interrupt these modern networks must use subtle, less obvious methodology that disguises the technique used, protecting the ability whenever possible to permit future use." In a similar solicitation in April 2007, the Air Force's 950th Electronic Systems Group said it wanted industry help to define technologies to "disrupt, deny, degrade, or deceive an adversary's information system." The service also seeks tools that will help it map and access data and voice networks, conduct denial-of-service attacks, and manipulate data on enemy networks.[225]

---

[223] Ibid.

[224] Coleman, K. (2007, November). Department of Cyber Defense: An organization whose time has come! The Technolytics Institute, 7 pp.

[225] Brewin, B. (2007, October 24). Management matters: Cyber wars. *GovernmentExecutive.com.* Retrieved May 3, 2008, from *http://www. govexec.com/story_page_pf.cfm?articleid=38352&printerfriendlyvers=1.*

Central Organization in the DoD to Fight Cyberwarfare

Recommendation No. 3. *There must be a clear central organization in the DoD that is trained and equipped to fight cyberwarfare and respond to large-scale foreign attacks by governments, terrorists, and extremists.*

The Threat. Malicious e-mail and other cyber attacks on Tibet advocacy groups in the United States are linked to Internet servers used in past hacker intrusions that U.S. law enforcement traced to China. Based on publicly available data, the link is the first direct evidence that the recently intensified attacks against the Tibet groups were launched from China. These attacks were reported by *United Press International* in March 2008. However, it is unclear whether or to what extent the Chinese government or military were involved.[226] With this threat, the United States must establish a clear, central organization in the DoD to fight a cyber war. The ensuing paragraphs indicate that various organizations in the United States are attempting to do just that.

Air Force Cyber Command. With U.S. civil and military officials increasingly concerned about cyber attacks against American networks, the USAF is planning to establish what will probably be the largest and most comprehensive military organization to defend against cyber attack. Additionally, unlike the apparent efforts of the other U.S. military services in this field, the Air Force will conduct offensive cyberwarfare.[227]

The massive Air Force effort will pull together existing cyber-related units and establish new ones, all under the Air Force Cyber Command—AFCYBER in milspeak—and its operating arm, the 24th Air Force. According to Major General William T. Lord, the provisional commander of AFCYBER, the command and 24th Air Force will

---

[226] Waterman, S. (2008, March 24). Cyber-attacks on Tibet groups tied to China. *The Washington Times*. Retrieved May 4, 2008, from *http:// homelandsecurity.osu.edu/focusareas/cyberterrorism.html*.

[227] Polmar, N. (2008, March 25). Cyber defense—and attack. *Military. com*. Retrieved May 3, 2008, from *http://www.military.com/ forums/0,15240,164702,00.html*.

achieve "initial operational capability" on 1 October 2008. However, many components of the command are already operational.[228]

Federal Bureau of Investigation (FBI). In the summer of 2007, the FBI quietly established a task force involving U.S. intelligence and other agencies to identify and respond to cyber-threats against the United States. Called the National Cyber Investigative Joint Task Force, the group has "several dozen" personnel working together at an undisclosed location in the Washington area, said Shawn Henry, the FBI's deputy assistant director of its cyber-division. The task force looks at "all cyber-threats," he said, but is focused on "organizations that are targeting U.S. infrastructure."[229]

Homeland Security Department (HSD). The HSD has launched an overhaul of the government's computer security efforts "almost . . . like a Manhattan Project"in response to concerns that the nation's Internet system is vulnerable to hackers and online terrorists, Secretary Michael Chertoff said on April 8, 2008. "The time has come to take a quantum leap forward, to really engage in what I'd call a game-changer in how we deal with (cyber) attacks," Chertoff said.[230]

Cyber Defense Post to Entrepreneur. The Bush administration tapped a Silicon Valley entrepreneur to head a new inter-agency group charged with coordinating the federal government's efforts to protect its computer networks from organized cyber attacks. The White House selected Rod A. Beckstrom as a top-level adviser based in the DHS. Beckstrom is an author and entrepreneur best known for starting Twiki. net, a company that provides collaboration software for businesses. Beckstrom reports directly to DHS Secretary Michael Chertoff.[231]

---

[228] Ibid.

[229] Waterman, S. (2008, April 21). FBI organizes defense against cyber-attacks. *The Washington Times*. Retrieved May 4, 2008, from *http:// homelandsecurity.osu.edu/focusareas/cyberterrorism.html.*

[230] Keefe, B. (2008, April 8). Government trying to improve internet security. *Cox News Service*. Retrieved May 4, 2008, from *http://homelandsecurity. osu.edu/focusareas/cyberterrorism.html.*

[231] Krebs, B. (2008, March 19). White House taps tech entrepreneur for cyber defense post. *Washingtonpost.com*. Retrieved May 4, 2008, from *http:// homelandsecurity.osu.edu/focusareas/cyberterrorism.html.*

Beckstrom's candidacy was backed chiefly by top brass at the DoD and the NSA. However, Beckstrom's appointment raised a number of questions. James Lewis, director of technology and public policy for the Center for Strategic and International Studies, noted that DHS only recently had appointed Greg Garcia as assistant secretary for cyber security and telecommunications. Garcia is former head of the Information Technology Association of America. Lawmakers on Capitol Hill, who believed DHS was not placing a strong enough emphasis on cyber, fought for and won this position through tireless lobbying.[232]

Garcia, in turn, answers to Robert D. Jamison, who serves as Under Secretary for National Protection and Programs Directorate. When asked at a press briefing about a simulated cyber attack against the United States who would lead the government's response in the event of a sustained cyber attack on the federal government, Jamison said that duty would fall to him.[233]

Department of Cyber Defense. From the previous paragraphs, it is apparent that no coordinated approach exists to establish a central organization in the DoD to fight a cyber war. This is why this recommendation is a most important one in dealing with cyber threats and warfare. To be successful, we must create a single, strong, focused organization to execute decisive, successful cyberwarfare. Kevin Coleman wrote that a Department of Cyber Defense is an organization whose time has come. He suggested that the United States create and empower an organization to work with business, industry, and government agencies in a collaborative manner that allows for the defense of information assets owned by or operating within the United States. This organization should be responsible for coordinating defensive capacity across business, industry, and government. In addition, they would coordinate offensive capabilities across the multiple organizations within the DoD.[234]

---

[232] Ibid.

[233] Ibid.

[234] Coleman, K. (2007, November). Department of Cyber Defense: An organization whose time has come! The Technolytics Institute, 7 pp.

A Clear Response Doctrine

Recommendation No. 4. *The U.S. government needs to develop a clear response doctrine.*

This is a most important recommendation because even if we possess the will to take the offensive, own the tools of warfare, and maintain a focused central organization, we must have a clear response doctrine. We must maintain the goals/objectives, strategies/tactics, plans of action, and guiding principles for winning a cyber war.

National Strategy to Secure Cyberspace. Our *National Strategy to Secure Cyberspace* is part of our overall effort to protect the United States. It is an implementing component of the *National Strategy for Homeland Security* and is complemented by a *National Strategy for the Physical Protection of Critical Infrastructures and Key Assets*. The purpose of this document is to engage and empower Americans to secure the portions of cyberspace that they own, operate, control, or with which they interact. Securing cyberspace is a difficult strategic challenge that requires coordinated and focused effort from our entire society—the federal government, state and local governments, the private sector, and the American people.[235]

Consistent with the *National Strategy for Homeland Security*, the strategic objectives of this *National Strategy to Secure Cyberspace* are to:[236]

- Prevent cyber attacks against America's critical infrastructures (CIs)
- Reduce national vulnerability to cyber attacks
- Minimize damage and recovery time from cyber attacks that do occur

The *National Strategy to Secure Cyberspace* identifies eight major actions and initiatives for cyberspace security response:[237]

---

[235] Anonymous (2003, February). *The National Strategy to Secure Cyberspace*. Washington, DC: The White House.

[236] Ibid.

[237] Ibid.

1. Establish a public-private architecture for responding to national-level cyber incidents
2. Provide for the development of tactical and strategic analysis of cyber attacks and vulnerability assessments
3. Encourage the development of a private sector capability to share a synoptic view of the health of cyberspace
4. Expand the Cyber Warning and Information Network to support the role of DHS in coordinating crisis management for cyberspace security
5. Improve national incident management
6. Coordinate processes for voluntary participation in the development of national public-private continuity and contingency plans
7. Exercise cyber security continuity plans for federal systems
8. Improve and enhance public-private information sharing involving cyber attacks, threats, and vulnerabilities

National Strategy Needs More Clarity. The *National Strategy to Secure Cyberspace* states that the private sector now has a crucial role in protecting national security because it largely runs the nation's CI. Tightly coupling business and industry into the cyber war defense strategy is arguably the most critical component. It represents the one area that the government has the worst track record, which must be improved. In addition, if we are to protect our nation, the *National Strategy to Secure Cyberspace* must contain language that requires by law that business, industry, and government adopt a set of minimal cyber security measures to protect our nation's information assets.[238]

2006 National Military Strategy for Cyberspace Operations (NMSCO). Cyberspace is also not confined only to the Internet. A presentation by Dr. Lani Kass titled "Cyberspace: A Warfighting Domain" cites the classified NMSCO defines *cyberspace* as a domain characterized by the use of electronics and the electromagnetic spectrum to store, modify, and exchange data via networked systems and associated physical infrastructures. The NMSCO also

---

[238] Coleman, K. (2007, November). Department of Cyber Defense: An organization whose time has come! The Technolytics Institute, 7 pp.

states that as a war-fighting domain . . . cyberspace favors the offense. Offensive capabilities in cyberspace offer both the United States and our adversaries an opportunity to gain and maintain the initiative.[239]

Right to Respond Unilaterally

Recommendation No. 5. *The United States should reserve the right to respond unilaterally to attacks against its infrastructure.*

To close out our five most important recommendations, for our survival, we must always reserve the right to respond unilaterally to attacks against our CI. This recommendation is extremely important for our survival as a country. Whatever it takes to vanquish those who chose to attack our CI, we must do first or subject ourselves to annihilation or bondage.

The Precedence Has Been Set. President Bush said that terrorism cells in countries that make up close to one-third of the globe must be actively sought and dismantled. "We must take that battle to the enemy, disrupt his plans, and confront the worst threats before they emerge," he said, adding that Americans must be "ready for pre-emptive action when necessary to defend our liberty and to defend our lives." He further said, "In the world we have entered, the only path to safety is the path of action. And this nation will act." President Bush took the Global War on Terrorism (GWOT) to Iraq in a preemptive first strike on Saddam Hussein's regime. That set the precedence for a first strike cyber attack on any country (such as Russia and China) bent on cyberwarfare.[240]

Offensive Strategies to Attack Enemy Cyber Assets. The Pentagon's network warriors have traditionally focused on defense. Not anymore. There's been a "radical change in U.S. policy when it comes to its cyber war-fighting stance," according to *Inside the Air Force.* A

---

[239] Kass, L. (2006, September 26). Cyberspace: A warfighting domain. AF Cyberspace Task Force PowerPoint presentation charts.

[240] Doran, J. (2002, June 3). Terror war must target 60 nations, says Bush. *Times Online.* Retrieved May 6, 2008, from *http://mprofaca.cro.net/ crimorder.html.*

few months back, Pentagon officials were saying that "the military had no plans to shift its cyberwarfare focus from a defensive mindset to an offensive one." But now that the Air Force has declared themselves the service in charge of all things electronic, "high-ranking service officials say they are developing offensive strategies to attack enemies' cyber assets."[241]

Cyber Favors the Offense. "Cyber, as a warfighting domain . . . like air, favors the offense," said Lani Kass, director of the Air Force's Cyberspace Task Force, while presenting the elements of the NMSCO in September 2007. This is the first time the Air Force formally acknowledged it has plans to take an offensive approach to cyberwarfare. The first battle of the next war will be fought and won in the cyberspace arena, she insisted.[242]

China's First-strike Capabilities. China's military has developed cyberwarfare first-strike capabilities that include units charged with developing viruses to attack enemy computer networks, a DoD report warned in May 2007. "The PLA [People's Liberation Army] has established information warfare units to develop viruses to attack enemy computer systems and networks, and tactics and measures to protect friendly computer systems and networks," the Pentagon's annual report to Congress on China's military power said. "In 2005, the PLA began to incorporate offensive CNO [computer network operations] into its exercises, primarily in first strikes against enemy networks."[243]

---

[241] Shachtman, N. (2007, October 9). Air Force readying cyber strikes. *Wired. com.* Retrieved May 3, 2008, from *http://blog.wired.com/defense/2007/10/also-nsa-target.html.*

[242] Reed, J. (2007, October 5). As effort ramps up …: Officials announce Cyber Command will take an offensive posture. *Inside the Air Force.* Retrieved May 3, 2008, from *http://integrator.hanscom.af.mil/2007/October/10112007/10112007-14.htm.*

[243] Keizer, G. (2007, May 29). China makes viruses for cyberwar first-strike. *Computerworld.com.* Retrieved May 6, 2008, from *http://www.computerworld.com/action/article.do?command=viewArticleBasic&articleId=9021663.*

Conclusion

One thing is for sure: success in future conflicts will depend less on bombs and bullets and more on bits and bytes.[244] In the end, the cyber threat is revolutionary, officials said, because it has no battle lines, the intelligence is intangible, and attacks come without warning leaving no time to prepare defenses. Education and training of computer users, not enforcement, are the most effective defense measures, officials said.[245]

If our enemies conduct a cyber war against us, we must be prepared to defend ourselves and also be capable of taking the offensive. *The best defense is a good offense.* Little is known about the precise nature of Washington's offensive capabilities. State—or group-sponsored attacks against our information systems using computer viruses and other techniques should be considered an act of war. Tightly coupling business and industry into the cyber war defense strategy is arguably the most critical component. It represents the one area that the government has the worst track record, which must be improved. As a war-fighting domain, cyberspace favors the offense. The first battle of the next war will be fought and won in the cyberspace arena.

---

[244] Coleman, K. (2007, November). Department of Cyber Defense: An organization whose time has come! The Technolytics Institute, 7 pp.

[245] Rogin, J. (2007, February 13). Cyber officials: Chinese hackers attack 'anything and everything.' *FCW.com*. Retrieved May 3, 2008, from *http:// www.fcw.com/online/news/97658-1.html?type=pf.*

Chapter 8
Strategy to Combat Cyberterrorism

Robert T. Uda, MBA, MS

This chapter covers the following five points of a proposed congressional strategy to combat cyberterrorism:

- What is *Cyberterrorism*?
- Risks of an Attack to Our Nation's Public Entities and Private Sector
- Measures That Can be Implemented to Address the Threat
- Problems with Past Government Programs
- Policy Recommendations That Would be Beneficial and Associated Reasoning

The ensuing sections detail each of these five points to assist the Congressional House Committee on Cyber-Threats and Infrastructure Protection in developing a strategy to combat cyberterrorism.

What is Cyberterrorism?

One thing for sure, there is no uniform consensus on a universal definition of the word *cyberterrorism*. Do you wonder why there is difficulty in obtaining international agreement by what we mean when we speak of *cyberterrorism*? In the following paragraphs, we list several definitions found in the open literature.

Definition #1. Once the terrorists have gained control of the system, they can abuse it in such a way as to cause major damage to human life and the government. Their actions, thereby, create major economic disruption. To cause this harm, it is not necessary for the terrorists to be physically co-located within the system facilities or even within the United States. This type of terrorist behavior is called *cyberterrorism.*[246]

Definition #2. *Cyberterrorism* is the malicious conduct in cyberspace to commit or threaten to commit acts dangerous to human life. Concurrently, these acts may be against a nation's critical infrastructure (CI) such as energy, transportation, or government operations. Terrorists commit these acts in order to intimidate or coerce a government or civilian population, or any sequence thereof, in furtherance of political or social objectives.[247] What is interesting here is that definitions #1 and #2 come from the same authors and the same book! They could not even agree on a single definition for *cyberterrorism* that they use in the same book.

Definition #3. *Cyberterrorism* (effects-based) exists when computer attacks result in effects that are disruptive enough to generate fear comparable to a traditional act of terrorism, even if done by criminals.[248]

Definition #4. *Cyberterrorism* (intent-based) exists when unlawful or politically motivated computer attacks are conducted to intimidate or coerce a government or people to further a political objective, or to cause grave harm or severe economic damage.[249],[250]

Definition #5. *Cyberterrorism* is terrorism that involves computers, networks, and the information they contain.[251]

---

[246] Bullock, J. A., Haddow, G. D., Coppola, D., Ergin, E., Westerman, L., & Yeletaysi, S. (2006). *Introduction to Homeland Security, Second Edition.* Oxford, United Kingdom: Elsevier Butterworth-Heinemann.

[247] Ibid.

[248] Rollins, J., & Wilson, C. (2007, January 22). Terrorist capabilities for cyberattack: Overview and policy issues. Congressional Research Service (CRS) Report for Congress, Order Code RL33123.

[249] Wilson, C. (n.d.). Computer Attack and Cyberterrorism: Vulnerabilities and Policy Issues for Congress. CRS Report RL32114. This version was cited in Rollins & Wilson (2007).

[250] Rollins, J., & Wilson, C. (2007, January 22). Terrorist capabilities for cyberattack: Overview and policy issues. Congressional Research Service (CRS) Report for Congress, Order Code RL33123.

[251] Cereijo, M. (2006, May 9). Cyberterrorism. Retrieved March 22, 2008, from *http://www.canf.org/2006/1in/ensayos/2006-may-09-cyberterrorism.htm.*

Definition #6. *Cyberterrorism* involves the use of computer systems to carry out terrorist acts, which are, in turn, defined by reference to specific criminal statutes. True *cyberterrorism* is characterized by large-scale destruction (or the threat of such destruction) coupled with an intent to harm or coerce a civilian population or government.[252]

Definition #7. According to the U.S. Federal Bureau of Investigation (FBI), *cyberterrorism* is any "premeditated, politically-motivated attack against information, computer systems, computer programs, and data which result in violence against non-combatant targets by sub-national groups or clandestine agents."[253]

Definition #8. According to the U.S. National Infrastructure Protection Center (NIPC), *cyberterrorism* is a criminal act perpetrated by the use of computers and telecommunications capabilities. This act results in violence, destruction, and/or disruption of services to create fear by causing confusion and uncertainty within a given population. All of this is done with the goal of influencing a government or population to conform to a particular political, social, or ideological agenda.[254],[255]

Conclusion. There you go. There are many, many more definitions for *cyberterrorism* that you can find in the general literature. Like its predecessor, *terrorism*, we would be hard pressed to develop a definition that every nation in the UN would agree upon. It would be

---

[252] Malcolm, J. (2004, February 24). Virtual threat, real terror: Cyberterrorism in the 21st century. Testimony to the United States Senate Committee on the Judiciary. Retrieved March 22, 2008, from *http://www.globalsecurity. org/security/library/congress/2004_h/040224-malcolm.htm.*

[253] Pollitt, M. M. (n.d.). *Cyberterrorism—Fact or Fancy?* Washington, DC: FBI Laboratory. Retrieved April 5, 2008, from *http://www.cs.georgetown. edu/~denning/infosec/pollitt.html.*

[254] Garrison, L., & Grand, M. (ed., 2001). Cyberterrorism: An evolving concept. *NIPC Highlights.* Retrieved from *http://www.nipc.gov/ publications/highlights/2001/highlight-01-06.htm.*

[255] Kerr, K. (2004, October 9). Putting cyberterrorism into context. Computer Crime Research Center, Source: AusCERT. Retrieved May 16, 2008, from *http://www.crime-research.org/articles/putting_cyberterrorism.*

an exercise in futility. At any rate, we move forward in our analysis of *cyberterrorism*.

Risks of an Attack to Our Nation's Public Entities and Private Sector

Cyberterrorism Risks. The roots of the notion of cyberterrorism can be traced back to the early 1990s. The rapid growth in Internet use and the debate on the emerging "information society" sparked several studies on the potential risks faced by the highly networked, high-tech-dependent United States. As early as 1990, the National Academy of Sciences began a report on computer security. They used the words: "We are at risk. Increasingly, America depends on computers. . . . Tomorrow's terrorist may be able to do more damage with a keyboard than with a bomb." At the same time, the prototypical term "electronic Pearl Harbor"was coined, which linked the threat of a computer attack to an American historical trauma.[256]

Risk Assessment. The first and most crucial step is to perform an accurate risk assessment. Risk is a variable that cannot be applied equally to every threat. Each has a different level of risk dependent on a broad range of circumstances, factors, and variables. Hence, a different level of protection is required for each threat.[257]

Information at Risk. The current state of cyberspace is such that information is seriously at risk. The impact of this risk to the physical health of mankind is, at present, indirect. At present, computers do not control sufficient physical processes (without human intervention) to pose a significant risk of terrorism in the classic sense. Therein rests the following two lessons:[258]

---

[256] Weimann, G. (2004, December). Cyberterrorism: How real is the threat? United States Institute of Peace Special Report No. 119. Retrieved April 4, 2008, from *http://www.usip.org/pubs/specialreports/sr119.html*.

[257] Ashenden, D. (2003, January). Protect and survive: Communication networks can only be protected against cyberterrorist attacks if telcos, governments, and end users work together. *Telecommunications International, 37(1)*, pp. 29-31.

[258] Pollitt, M. M. (n.d.). *Cyberterrorism—Fact or Fancy?* Washington, DC: FBI Laboratory. Retrieved April 5, 2008, from *http://www.cs.georgetown. edu/~denning/infosec/pollitt.html*.

- The definition of terrorism needs to address the fundamental infrastructure upon which civilization is increasingly dependent.
- A proactive approach to protecting the information infrastructure is necessary to prevent its becoming a more serious vulnerability.

As we build more and more technology into our civilization, we must ensure that there is sufficient human oversight and intervention to safeguard those whom the technology serves.

Cyberterrorist Attack. Michael Vatis, director of the Institute for Security Technology Studies (ISTS) at Dartmouth College works to identify top vulnerabilities. To the question, do you expect a cyberterrorist attack, he said, "Given that there has been evidence of Al-Qaeda planning for it and there still are Al-Qaeda members on the loose, I think we definitely could see direct cyberterrorist attacks. Professionally, I think a stand-alone cyber-attack is the most likely, rather than a coordinated effort with a physical attack. It definitely could be coupled with a physical attack, but it's easier to plan and execute a cyber-attack than plan the timing of a physical and cyber-attack."[259]

Kinds of Cyber-attacks. Cyber-attacks can be malicious or accidental. They can involve attacks by other nation states, organized groups, or individuals. They can be motivated by monetary gain, ill will, or political interests. Cyber-attacks can be directed at governments, firms, or individuals. Cyber-attacks can involve the theft or destruction of information, the theft of services or financial assets, or the destruction of hardware or software infrastructure. Cyber-attacks can result in financial loss, business or service interruption, or infrastructure destruction. Cyber-attacks can be aimed directly at disrupting business or government services or can be launched in conjunction with physical attacks in order to magnify effects or prevent effective response. Developing effective law enforcement or national security policies to deal with cyber threats is a national priority.[260]

---

[259] Gaudin, S. (2002, July 19). Security expert: US companies unprepared for cyberterror. *IT Management*. Retrieved November 25, 2002, from *http://itmanagement.earthweb.com/secu/print.php/1429851*.

[260] Anonymous (n.d.). The program on telecommunications and cyber security policy. The Global Information Society Project. Retrieved from *http://www.global-info-society.org/PLENSIA/plensia.pdf*.

Threats and Attackers. Security threats and attackers are turning professional. Network managers still need to stop the script-kiddies from defacing their websites, but it is becoming increasingly important to stop the professionals who want to steal valuable information. The new attackers search for vulnerabilities in the application and exploit these weaknesses. Attackers are bypassing die-traditional network, layer firewall, and intrusion detection system (IDS) defenses. Their exploits appear as legitimate traffic to the network layer defense. However, hiding in the application layer are deadly attacks.[261]

Working to counter these threats includes a number of security vendors—some established and some new. Well-known security vendors such as Cisco and Radware have been joined by new security players such as NetContinuum, Imperva, Citrix, Breach Security, Protegrity, and ConSentry Networks. Additional players are expected to join the field. Their goal is to take security to the next level and protect both the network and applications.[262]

The new application threats come primarily from three areas:[263]

- Viruses, worms, malware, and rootkits, i.e., malware that hides in the operating system's (OS's) kernel
- Attackers exploiting Web application vulnerability
- Internal users gone bad or external attackers stealing valuable data

Viruses and worms have been around for awhile. Malicious programs—malware—have become increasingly popular, and rootkits are a new threat. The problem is that all these threats hide in the application payload, while they bypass traditional network security. Nevertheless, enterprise managers need to ensure that these attacks are not spread by the network.[264]

A Cyber Jihad Going on Right Now. John Arquilla, associate professor of defense analysis at the Naval Postgraduate School, is an

---

[261] Layland, R. (2006, September 29). Application security: Countering the professionals. *RedOrbit NEWS*. Retrieved April 7, 2008, from *http://www. redorbit.com/modules/news/tools.php?tool=print&id=674569*.

[262] Ibid.

[263] Ibid.

[264] Ibid.

expert on unconventional warfare. He said, "When we think about Al-Qaeda and its potential for cyberterror or other sympathetic Muslim groups, we're now in an area that's very proprietary in nature. All I can say on this subject is that there is a cyber jihad going on right now against Israel. We see some people that we associate with modern terrorism who are trying to use [a] cyberspace-based means to pursue their ends. Beyond that, I'm afraid we're in a much classified area."[265]

Reliance on Cyberspace Makes U.S. Uniquely Vulnerable. Lani Kass, director of the Air Force's Cyberspace Task Force, spoke at an Air Force Association-sponsored conference in Washington, DC, in September 2006. She said that "Groups like al-Qaeda and other extremist organizations can be effective using cyberspace"because "as a warfighting domain, it's different than the land, air, and space domains." In the symmetrical domain, we use expensive weapon systems like fighters, bombers, advanced ground vehicles, or aircraft carriers. However, in the cyberspace domain, everything one needs "[to] cause chaos from afar very cheaply . . . is available off the shelf," she said at the conference. Air Force leaders want to beef up the service's ability to guard against Internet-based attacks. The reason is because the United States "is uniquely vulnerable because of our reliance on cyberspace," both militarily and "in our everyday lives," she said. Cyberspace offers advantages to those who do not want to deal with U.S. forces in a symmetric fight, Kass added.[266]

Measures to Implement to Address the Threat

Dartmouth College Recommendations. One of the most timely and comprehensive studies on the subject is "Cyber-attacks During the War on Terrorism," prepared by the Dartmouth College Institute for Security

---

[265] Arquilla, J. (2003, April 24). Cyber war!: Interviews: John Arquilla. *Frontline*, Public Broadcasting System. Retrieved May 3, 2008, from *http://www.pbs.org/wgbh/pages/frontline/shows/cyberwar/interviews/ arquilla.html.*

[266] Bennett, J. T. (2006, October 4). Air Force to establish new Cyberspace Operations Command. *World Politics Review*. Retrieved May 6, 2008, from *http://www.worldpoliticsreview.com/articlePrint.aspx?ID=233.*

Technology Studies. It was published just 11 days after the Trade Center disaster. Among its recommendations included the following:[267]

- Operating systems and software should be updated regularly.
- Strong password policies should be enforced.
- Systems should be "locked down" whenever possible.
- Anti-virus software should be kept up-to-date.
- High fidelity intrusion detection systems and firewalls should be employed.
- All vital data should be backed up regularly and stored off-site to prevent loss in the case of a physical or cyber-attack.
- All the measures to secure critical infrastructure assets should be clearly explained in an enforceable security policy.

The Dartmouth study emphasizes that security measures previously considered excessive should now be considered a minimum effort. Since 2002, most government agencies and private sector companies (primarily high-tech businesses) implement all or most of these recommendations to some degree or other. We need to do more and strive to be better at it.

GAO Recommendations. Another analysis conducted in 2003 by the U.S. Government Accountability Office (GAO) found significant information security weaknesses at 24 major government agencies. The GAO report said, "Further information security improvement efforts are needed at the government-wide level." "These efforts need to be guided by a comprehensive strategy in which roles and responsibilities are clearly delineated, appropriate guidance is given, adequate technical expertise is obtained, and sufficient agency information security resources are allocated."[268]

---

[267] Harper, D. (2002, January). Cyberterror: A fact of life: Experts anticipate a rise in terrorist attacks on and through computer systems. *Industrial Distribution, 91(1)*, p. 68. Retrieved May 15, 2008, from *http://www.accessmylibrary.com/coms2/summary_0286-24952945_ITM.*

[268] Rothman, P. (2003, May 1). How can we protect our critical infrastructure from cyber-attack? *Government Security.* Retrieved April 24, 2008, from *http://govtsecurity.com/mag/protect_critical_infrastructure/.*

The GAO identified several areas of weakness among the systems and issued the following recommendations:[269]

- Develop a comprehensive and coordinated national critical infrastructure protection (CIP) plan
- Improve information sharing on threats and vulnerabilities both among government agencies and between the private sector and the federal government
- Improve analysis and warning capabilities for both cyber and physical threats
- Encourage entities outside the federal government to increase their CIP efforts

Since 2003, all of these recommendations are being implemented in varying degrees by many agencies of the federal government and companies within the private sector. Again, we just need to do more and strive to be better at it.

Private Sector is Key. The private sector must undertake most of the responsibility for fixing weaknesses in key Internet assets. Business executives are dependent on a patchwork of public—and private-response programs to restore Internet infrastructure services. In many cases, these programs are not fully coordinated via a central organization. Immediate—and long-term commitments to change the current reality should include the following steps:[270]

- Establish a single point of contact (POC) and responsibility for government interaction
- Set strategic needs and direction
- Consolidate early warning and response organizations
- Agree on an information-sharing mechanism

---

[269] Ibid.
[270] Anonymous (2006, June). Essential steps to strengthen America's cyberterrorism preparedness: New priorities and commitments from Business Roundtable's Security Task Force. Business Roundtable, 24 pp. Retrieved from *http://www.businessroundtable.org/pdf/20060622002Cyb erReconFinal6106.pdf.*

We need the private sector to be fully engaged in these activities.

Adversarial Neutrality Required. In recent congressional testimony, Director of National Intelligence Michael McConnell named Russia and China as among the most important cyber-adversaries of the United States. Shawn Henry, the FBI's deputy assistant director of its cyber-division, said it is important to be "adversary neutral"in combating cyber-threats. "A network can be attacked by a terrorist group, a foreign power, or a hacker kid from Oklahoma City. Networks need to be protected from all threats because once [sensitive] data has been stolen, it can be transferred anywhere," he said. In recent testimony, Mr. McConnell said the U.S. government is "not prepared to deal with" the cyber-threats it faces. Additionally, Homeland Security Secretary Michael Chertoff told a bloggers roundtable in March 2008 that cyber security is "the one area in which I feel we've been behind where I would like to be."[271] Hence, we should protect simultaneously against cybercrime, cyberterrorism, and cyberwarfare. We must be adversary neutral and prepare for all adversaries whether criminals, terrorists, and/or cyber-warriors of enemy nations.

Asked whether the U.S. government is resolving the problem, Mr. Henry said, "Our response has to change constantly and grow because the threat is constantly changing and growing." He said that one of the most worrisome aspects of cyber-threats is the extent to which "the offense outstrips the defense." "The pace of technological change—the increasing connectivity of networks—creates more opportunity for exploitation" of vulnerabilities, he said.[272] It is obvious that we must continue our R&D and develop new countermeasures to defend against new, more sophisticated cyber-threats. Our enemies only need to be successful once; we need to be successful 100 percent of the time.

Problems with Past Government Programs

Three Cyber-gaps. Many cyber security observers are concerned that U.S. government efforts to date have not effectively prepared the nation for a catastrophic cyber-attack. A Business Roundtable

---

[271] Waterman, S. (2008, April 21). FBI organizes defense against cyber-attacks. *The Washington Times*. Retrieved May 4, 2008, from *http:// homelandsecurity.osu.edu/focusareas/cyberterrorism.html.*

[272] Ibid.

report issued in June 2006 found three "cyber-gaps" that are keeping the United States from being prepared to recognize and respond to a cyber-attack:[273],[274]

- *Indicators*—The lack of established indicators that would indicate an attack is underway
- *Responsibility*—A failure to identify who is responsible for restoring affected infrastructure
- *Resources*—A lack of dedicated resources to assist in returning cyber operations to a pre-attack condition

Evidence in the general literature indicates that these gaps are currently being addressed by the government and private sector. These evidences include installed alert and warning systems, published strategies and plans, and increased funding for countering cyber-threats.

A Failure to Communicate. One of the most difficult challenges facing the field of CIP is the lack of shared terminology. There are too many people using too many ill-defined terms for the community of homeland security experts to communicate properly. The lack of widely accepted definitions of terms used in homeland security leads to reinvention of the wheel, false starts, and more detours.[275] Perfect examples are the existence of myriad definitions of *terrorism* and *cyberterrorism* . . . among others. Obviously, a need exists for a standardization organization, industry standards, and common practices for the homeland security industry. We need organizations similar to the following . . . but for homeland security:

---

[273] Anonymous (2006, June). Essential steps to strengthen America's cyberterrorism preparedness: New priorities and commitments from Business Roundtable's Security Task Force. Business Roundtable, 24 pp. Retrieved from *http://www.businessroundtable.org/pdf/20060622002 CyberReconFinal6106.pdf.*

[274] Rollins, J., & Wilson, C. (2007, January 22). Terrorist capabilities for cyberattack: Overview and policy issues. Congressional Research Service (CRS) Report for Congress, Order Code RL33123.

[275] Lewis, T. G., & Darken, R. (2005). Potholes and detours in the road to critical infrastructure protection policy. *Homeland Security Affairs Journal, I*(2), Article 1. Retrieved from *http://www.hsaj.org/hsa/vol1/iss2/art1.*

- International Organization for Standardization (ISO)
- International Telecommunication Union (ITU)
- American National Standards Institute (ANSI)
- American Society for Testing and Materials (ASTM)
- Institute of Electrical & Electronics Engineers (IEEE)
- Society of Automotive Engineers (SAE)
- Software Engineering Institute (SEI)

There are as many definitions of 'vulnerability'and 'risk'as there are agencies in federal, state, and local governments . . . combined! Before we can take the first step in a 1,000-mile journey, we need a compass. Currently, there is no universally accepted definition of the most basic measures of criticality—vulnerability and risk.[276]

Increasing Use of the Internet. Increased security measures are being applied to physical facilities. Additionally, increased efforts are being made by the U.S. government to track and engage groups in their home countries. Hence, many believe that the Internet will increasingly play a larger role in terrorist support and operational efforts. Many observers that monitor the Internet suggest that, due to the effects of intensified counterterrorism efforts worldwide, Islamic extremists are gravitating toward the Internet. Furthermore, they are succeeding in organizing online where they have been failing in the physical world. Terrorist groups increasingly use online services for covert messaging through steganography, anonymous e-mail accounts, and encryption.[277],[278] *Steganography* is a means of protecting data confidentiality by "hiding" it within a larger data file.

Our Defense Cannot Keep Up with the Offense. At a recent computer security conference In San Francisco in April 2008,

---

[276]   Ibid.

[277]   Tendler, S. (2005, July 20). Encrypted files frustrate police. *Times Online*. Retrieved from http://technology.timesonline.co.uk/article/0,20409-1701405,00.html, CryptoHeaven at http://www.cryptoheaven.com, and SecretMaker at *http://www.secretmaker.com/emailsecurer/steganography/default.html*.

[278]   Rollins, J., & Wilson, C. (2007, January 22). Terrorist capabilities for cyberattack: Overview and policy issues. Congressional Research Service (CRS) Report for Congress, Order Code RL33123.

Department of Homeland Security (DHS) Secretary Michael Chertoff said the government has taken some strides in making its computer networks more secure since his department was created in 2003. But with constantly evolving threats and computer networks becoming increasingly important, the government must take steps comparable to the World War II "Manhattan Project"effort to create the atomic bomb, he said.[279] Again, Congress sits on its collective gluteus waiting for a major cyber-attack to motivate them to action. By then, it will be too little too late.

In March 2008, a GAO report found that computer security problems are common—and growing—throughout federal agencies. According to the report, software used by government agencies contain as many as 29,000 security vulnerabilities that could allow a hacker to compromise government computers. Meanwhile, according to the GAO, the number of computer attacks and related incidents reported by government agencies has soared by nearly 260 percent over the years 2005 through 2007.[280] What is the Administration and Congress doing about these startling facts? Nothing much. In the past 40 years, we have never had such a do-nothing Congress as the current one in power.

Deficiencies of Government Systems. Although government systems may have deficiencies, a greater vulnerability may lie with critical infrastructures. Finance, utilities, and transportation systems are predominately managed by the private sector and are far more prone to attacks because those organizations are simply unprepared. A survey by the United Kingdom (UK)-based research firm, Datamonitor, shows that businesses have been massively under-spending for computer security. Datamonitor estimates that $15 billion is lost each year through E-security breaches, while global spending on defense is only $8.7 billion. Moreover, even if business were to improve its security spending habits and correct the weaknesses in its computer systems, it is impossible to eliminate all vulnerabilities. Administrators often ignore good security practices or are unaware of weaknesses when they

---

[279] Keefe, B. (2008, April 8). Government trying to improve internet security. *Cox News Service.* Retrieved May 4, 2008, from *http://homelandsecurity. osu.edu/focusareas/cyberterrorism.html.*

[280] Ibid.

configure systems. Furthermore, there is always the possibility that an insider with knowledge may be the attacker.[281]

Federal Grants. Improving the partnership among federal and nonfederal officials is vital to achieving important national goals. The task facing the nation is daunting and federal grants will be a central vehicle to improve and sustain preparedness in communities throughout the nation. While funding increases for combating terrorism have been dramatic, the GAO's report reflects concerns that many people maintain about the adequacy of current grant programs to address the homeland security needs.[282]

Ultimately, the "bottom line" question is this: What impact will the grant system make in protecting the nation and its communities against terrorism? At this time, it is difficult to know since we do not possess clearly-defined national standards or criteria defining existing or desired levels of preparedness across the country. Our grant structure is not well-suited to provide assurance that scarce federal funds are, in fact, enhancing the nation's preparedness in the places that are most at risk.[283] The lack of "clearly-defined national standards or criteria" again rears its ugly head.

There is a fundamental need to rethink the structure and design of assistance programs, streamline/simplify programs, improve targeting, and enhance accountability for results. Federal, state, and local governments alike own a stake in improving the grant system. This must be done to reduce burden and tensions and promote

---

[281] Gabrys, E. (2002, September/October). The international dimensions of cybercrime, part 1. *Information Systems Security, 11(4)*, 21-32. Retrieved April 7, 2008, from *http://firstsearch.oclc.org.proxy1.ncu.edu/images/WSPL/wsppdf1/HTML/06470/M3D8X/TSG.HTM.*

[282] Posner, P. L. (2003, September 3). Homeland security: Reforming federal grants to better meet outstanding needs. Statement of Paul L. Posner, managing director Federal Budget Issues and Intergovernmental Relations, Strategic Issues, before the Subcommittee on Terrorism, Technology, and Homeland Security, Committee on the Judiciary, US Senate. United States Government Accountability Office (GAO) Report Number GAO-03-1146T, 24 pp. Retrieved May 16, 2008, from *http://www.gao.gov/cgi-bin/getrpt?GAO-03-1146T.*

[283] Ibid.

the level of security that can only be achieved through effective partnerships. The sustainability and continued support for homeland security initiatives rest in no small part on all of us. It depends on our ability to demonstrate to the public that scarce public funds are in fact improving security in the most effective and efficient manner possible.[284]

$844 Million in CIP Grants for FY 2008. The DHS will distribute $844 million in FY 2008 in infrastructure protection grants for security at ports, trucking, bus systems, ferry systems, and other critical infrastructure (CI) facilities. This is a significant increase of nearly $189 million over 2007's amount in funding for security programs.[285]

These grants are part of the Administration's efforts to strengthen the security of the country's CI. Most of America's critical infrastructure is owned and/or operated by state, local, and private sector partners. DHS'Infrastructure Protection Activities grants are set aside for state and local agencies, port authorities, and owners and operators of transit systems.[286]

Since 2002, $3 billion has been spent on infrastructure protection grants. In prior years, the government used grants to fund response and recovery capabilities, capital equipment, and assets. Funding increases for FY 2008 are for prevention of improvised explosive devices (IEDs); information sharing; communication; and more regionally-based security cooperation and exercising.[287]

Policy Recommendations that would be Beneficial

National Policy Needed. The lack of a national policy on Internet reconstitution could undermine the economy and the security of the nation. The identified gaps and possible solutions do not require extensive funding. In addition, solution implementation does not require massive government reorganization. Instead, both the public and private sectors must commit and focus their efforts and funding on specific capabilities to put strategies and plans in place to reconstitute the Internet following

---

[284] Ibid.
[285] Anonymous (2008, May 28). $844 million in infrastructure protection grants. *Homeland Defense Journal Weekly Newsletter, Issue 22.*
[286] Ibid.
[287] Ibid.

a significant disruption. A coordinated response will help our nation and our economy recover quickly following a cyber-attack.[288]

Adopt Policies that Ensure Critical Government Services. Federal, state, and local governments have unique roles in ensuring vital government services—national defense, rule of law, and emergency services readiness—even under the stressful conditions of information warfare (IW) attack. Maintaining continuity in these areas can prove to be challenging and expensive. Government officials need to identify those functions that only government can perform and ensure that government has secure information systems and processes to maintain these functions. This approach requires updating and expanding government plans for the Information Age and securing the essential infrastructures upon which all levels of government depend.[289]

Director of National Intelligence (DNI). The ability exists of a catastrophic cyber-attack to disrupt a significant portion of the nation's infrastructure. Some national security observers suggest that the DNI should be responsible for monitoring these capabilities. Additionally, he should monitor the identities of the countries and groups that may wish to cause our nation harm through cyber-attacks. The DNI is our nation's chief intelligence officer. He possesses the ability to coordinate all known cyber-threat-related information. Then, he tasks the intelligence community to collect information to understand better the groups that may wish to cause the United States harm and to forecast their intentions and capabilities.[290]

Security Policy. A security policy also should specify the technologies and procedures that will provide the protection. Security

---

[288] Anonymous (2006, June). Essential steps to strengthen America's cyberterrorism preparedness: New priorities and commitments from Business Roundtable's Security Task Force. Business Roundtable, 24 pp. Retrieved from *http://www.businessroundtable.org/pdf/20060622002Cyb erReconFinal6106.pdf.*

[289] Anonymous (2001-2002). Cyberterrorism and cyberwarfare thus become a plausible alternative: Summary of recommendations. Computer Crime Research Center (CCRC). Retrieved May 16, 2008, from *http://www. crime-research.org/library/Judge3.htm.*

[290] Rollins, J., & Wilson, C. (2007, January 22). Terrorist capabilities for cyberattack: Overview and policy issues. Congressional Research Service (CRS) Report for Congress, Order Code RL33123.

likely will include antivirus software, firewalls, intrusion detection, and data backup systems. The policy also should include details such as how employees identify themselves and how often they must change passwords.[291]

Ideal Policy Directions. Ideal policy directions have been nicely mapped out. These include:[292]

- **Laws.** The enactment of substantive and procedural laws, which are adequate to cope with current and anticipated manifestations of cybercrime
- **Forensics.** The development of forensic computing skills by law enforcement and investigative personnel and judicial officers
- **Legal Harmony.** The achievement of a modicum of legal harmonization (ideal at a global level)
- **Cooperation.** The creation of mechanisms for operational cooperation among law enforcement agencies from different countries—24/7 POCs for investigators and mechanisms for mutual assistance in cyber criminal matters generally.

Policy Adjustment Required. Terrorism is not only a criminal activity . . . it is a military assault on the entire population. Hence, we must disavow the notion that local law enforcement agencies are capable of preventing acts of violence against CI assets. An attack on the Weston building telecom hotel located in Seattle is not a criminal activity against Seattle, but a military action against the entire country. It must be dealt with as such.[293]

---

[291] Misra, S. (2003, June). High-tech terror: Cities and counties need plans to respond to criminal efforts to destroy government computer networks and data. *The American City & County, 118*(6), p. HS6.

[292] Bullwinkel, J. (2005). International cooperation in combating cybercrime in Asia: Existing mechanisms and new approaches. In R. Broadhurst, & P. Grabosky (Eds.), *Cybercrime: The challenge in Asia* (pp. 269-302). Hong Kong: Hong Kong University Press.

[293] Lewis, T. G., & Darken, R. (2005). Potholes and detours in the road to critical infrastructure protection policy. *Homeland Security Affairs Journal, 1(2)*, Article 1. Retrieved from *http://www.hsaj.org/hsa/vol1/iss2/art1*.

Issues for Congress. Policy issues for cybercrime and cyberterrorism include a need for the following:[294]

- **Threats.** Increase awareness about changing threats due to the growing technical skills of extremists and terrorist groups
- **Metrics.** Develop more accurate methods for measuring the effects of cybercrime
- **Response.** Help to determine appropriate responses by DoD to a cyber-attack
- **Incentives.** Examine the incentives for achieving the goals of the *National Strategy to Secure Cyberspace*
- **Software.** Search for ways to improve the security of commercial software products
- **Education.** Explore ways to increase security education and awareness for businesses and home personal computer (PC) users
- **Coordination.** Find ways for private industry and government to coordinate to protect against cyber-attack

Congress may also wish to consider ways to harmonize existing federal and state laws. These laws require notice to persons when their personal information has been affected by a computer security breach. Additionally, these laws impose obligations on businesses and owners of that restricted information.[295]

Conclusion

There is no uniform consensus on a universal definition of the word *cyberterrorism*. We would be hard pressed to develop a definition that every nation in the UN would agree upon. So, we move forward anyway.

Frank Cilluffo, an analyst at the Center for Strategic and International Studies in Washington, DC, testified to the Senate

---

[294] Wilson, C. (2008, January 29). Botnets, cybercrime, and cyberterrorism: Vulnerabilities and policy issues for Congress. Congressional Research Service (CRS) Report for Congress, Order Code RL32114, 43 pp.

[295] Ibid.

Government Affairs Committee in October 2001. He said, "Bits, bytes, bugs, and gas will never replace bullets and bombs as the terrorist weapon of choice," However, "while [Osama] bin Laden may have his finger on the trigger, his grandson may have his finger on the mouse."[296] Tomorrow's terrorist may be able to do more damage with a keyboard than with a bomb.[297]

Developing effective law enforcement or national security policies to deal with cyber threats is a national priority. However, the private sector must undertake most of the responsibility for fixing weaknesses in key Internet assets. We must understand that it is impossible to eliminate all vulnerabilities. Finally, terrorism is not only a criminal activity—it is a military assault on the entire population, and it must be dealt with accordingly.

---

[296] Verton, D. (2002, January 7). Critical infrastructure systems face threat of cyberattacks. *Computerworld, 36(2)*, p. 8. Retrieved May 15, 2008, from *http://www.computerworld.com/printthis/2002/0,4814,67135,00.html.*

[297] Weimann, G. (2004, December). Cyberterrorism: How real is the threat? United States Institute of Peace Special Report No. 119. Retrieved April 4, 2008, from *http://www.usip.org/pubs/specialreports/sr119.html.*

Chapter 9
Fighting Cybercrime, Cyberterrorism, and Cyberwarfare

Darrin L. Todd, MS

Ⅰt would be difficult to address the myriad of ways in which someone could commit cybercrimes, cyberterrorism, or cyberwarfare within the scope of one small chapter. In fact, there are numerous books that discuss a variety of technological devices and defensive techniques that may be employed to defend against cyber attacks. However, within the scope of this chapter, we can discuss a high-level or 'birds-eye' view of defending against various cyber attacks and attackers using some simple principles of information security process management.

Perspectives on Information Security

Though often erroneously thought to be synonymous with computer security, information security defines the broader scope of information asset protection. As with computer security, information security includes the core fundamental protective criteria of confidentiality, integrity, and availability as well as the expanded parameters of authenticity, accountability, and non-repudiation. Employing all of these criteria should ensure a comprehensive organizational security program.

The reality of information security is that it is a continuous process of persistently redefining and assessing risks, threats, and probable targets of cyber-attacks. The growing dependence upon information assets, the global nature of today's information security threats, the increasing numbers of attached systems and networks, and the rising complexity of system capabilities creates additional requirements for sound information security practices.

Information Security Process Management

Information security process management is a term that I've coined to get people out of the habit of believing that information security is a definable object that can be possessed or obtained. Rather, information security should be thought of as a perpetual process of steps taken to preserve or increase the security posture of an organization. It is a cyclical process; a never-ending procedure involving the constant review of attacks, attack trends, and the defensive measures and strategies deployed to counter them. Thinking of security as a continuously replenishing 'shopping list' versus a one-time purchase will provide the best climate for a strong defense against cyber-attacks.

Management of Information Security

There are a series of fundamental steps that may be taken to help 'secure' your network from a cyber-attacker; most of which do not involve yet another plug-in device to add more network latency and a false sense-of-security. Don't get me wrong. Technology certainly has its place within a good information security program, yet it is the exclusive reliance upon security devices that often creates a euphoric feeling among information security professionals and organizational managers alike. The belief that technology solutions alone will secure a network can be demonstrably shown to have led to some of the largest, most devastating security failures in modern times; leading to untold billions of dollars in losses to corporations. Information security is a management problem, not a technology problem.[298]

A sound information security program begins with the appointment of an information security manager. The information security manager serves to enumerate security threats, analyze them, and assess risks to the organization and organizational assets. However, information security entails much more than appointing a great information security manager and charging them with the security of the organization's information infrastructure. It involves a top-to-bottom organizational commitment to information security.

---

[298] Panko, R.R. (2003). *Corporate Computer and Network Security*. Upper Saddle River, NJ: Prentice Hall Publishing.

If the information security personnel in an organization do not possess the authority to enforce security policies, they cannot be expected to control the conduct of the security program. The actions of the organizational users cannot be controlled if the security program has no enforceability or "teeth." Enforcement is a basic principle of information security. In order to be effective, the organizational security policy must have real repercussions for those that fail to abide by its tenets.[299]

Many security programs are based upon an unrealistic notion that security is a black and white proposition. In fact, there are many shades of grey in-between. The organizational comfort level associated with the security of information systems relies heavily upon the information security manager's declaration that organizational information systems are secure. However, such an assurance is a placebo and does nothing for the security of the information assets. In order to secure organizational information assets, we must be willing to concede that there is no such thing as one hundred percent security, though that is the goal.

Organizational managers should never be led to believe information assets are one hundred percent secure. Your candor and assurance that policies and measures are constantly being reviewed, updated, and enforced is an acknowledgement that you are aware of the complexities of security and the unknown quantity and quality of the cyber-attacker. Managers should know that information security always carries some element of risk. It is the job of the information security manager to convey those risks to higher management. Some security risks may be minimized, managed, or controlled and others must be accepted as part of 'playing the game.'[300] What appeared to be relatively secure yesterday may not be secure today.

We can now discuss the basics of security that must be applied to ensure a foundation for the tight, consistent control of an information security program that provides for comprehensive protection against a wide variety of attackers and attack types. It's nice to discuss management concepts, yet it is the omnipresent and complex daily tasks of managing

---

[299] Pfleeger, C., & Pfleeger, S. (2003). *Security in Computing* (Third Edition). Upper Saddle River, NJ: Prentice Hall.

[300] Panko, R.R. (2003). *Corporate Computer and Network Security*. Upper Saddle River, NJ: Prentice Hall Publishing.

an information security program that presents the challenge. There is a continuous stream of security issues that information security managers must deal with while managing the threat of cyber-attacks.

Managing IT Facilities from a Security Viewpoint

Well-written and incorporated operations and contingency plans are essential to the safeguarding of information assets and preparing for operational emergencies. Safeguarding contains three essential elements; anticipating risks and threats, developing and employing countermeasures, and employing mechanisms to negate or minimize damage should preventative measures fail.

To effectively prepare IT facilities for risks and threats, information security managers must review every process and every detail from a security perspective. Nothing should be taken for granted. Effective risk management entails measurement of risks and their potential costs to the organization, should they become a reality. Spending excessive funds to secure a low-priority asset is not cost-effective and drains funds that may otherwise be used for other security priorities. As Goodman and Ramer[301] noted "Risk management must be cost effective. All vulnerabilities are not equal. An unlocked door is not a problem in a garden shed but if it leads to a bank vault or a data center that is another story."

Preparation for unforeseen circumstances entails meticulous requirements for the documentation of all routine and emergency procedures. All organizations should develop a crisis management plan that anticipates a wide range of scenarios. This plan should be heavily integrated with the organizational security policy and developed to conform to corporate security policies and directives. The IT facility security policy combines the corporate security philosophy and guidance, the specific requirements of the IT facility and the changing operating environment. The security policy is the element from which the IT facility reviews every process, task, and asset to determine and counter risks and threats. A sound security policy is the foundation of any successful security program.

---

[301] Goodman, S., & Ramer, R. (2007). Identify and mitigate the risks of global IT outsourcing. *Journal of Global Information Technology Management, 10*(4). Retrieved February 5, 2008, from ProQuest database.

Many security programs place emphasis on protections that guard against external threats with little regard for internal threats. There are many motives that can fuel an employee's desire to commit a criminal act that could jeopardize an information asset. Separation and rotation of duties, along with thorough system auditing and routine data backups can greatly assist with sabotage risks and threats.

Government organizations can differ from civilian organizations as the complexity of organizational intermingling can often create difficulties in assigning an information asset owner, though all government agencies maintain information security programs in full compliance with a set of similar security criteria. Government organizations are often targeted for different reasons by cyber criminals and are a prime target of espionage.

Malicious Software

Malicious software (virus software) has come to typify the concept of a security threat. There are many ways to minimize the threat of virus infections on a system and network. Organizations must educate employees about the hazard, including detection, defense, and reporting. Rigorous patch management and automated virus checking routines ensure the systems and networks are as resistant to known viruses as possible. Unauthorized software should be kept off the organizational network and Internet usage policies should be implemented; denoting what users can and cannot do using a corporate computer asset. As well, networks should be structured to contain or isolate virus outbreaks and minimize the damage to corporate systems.

Physical Security

Physical security is a component of information security and the overall security program. Physical security is the collection of protective strategies, controls, and devices that serve to disallow physical access to resources requiring protection. As with any security measure, physical protective measures are often a balancing act between allowing access to authorized personnel while disallowing access to unauthorized personnel. Physical security programs protect valuable assets, establish boundaries, conceal the nature of assets, protect information, provide for safety, prevent theft, and adapt to other situational criteria as required.

All security measures essentially begin with physical security. If you tightly control physical access to a device, you have taken a first step toward delaying or denying an attack. Physical security and physical asset control is the foundation upon which all other security measures are built. How will a high-technology solution prevent damage from an attack when an attacker gains access to the machine room and simply walks away with a server under-arm? If you deny an opportunity by locking and controlling access to the physical network components, you have taken a major step towards securing your network.

Current trends reveal that terrorists prefer physical attacks, though this could change. Cyber assets may be backed up by network defensive systems and measures, but physical attacks may create harm not only to networks and systems but to personnel and other assets within the facility being protected. Physical security measures are designed and implemented to minimize these vulnerabilities and risks, though some level of risk will always be present.

Some measures should be implemented to scrutinize personnel with authorized access to ensure their worthiness to retain access. The program should also provide for understanding of organizational policies for all personnel and provide user training in security and security response measures. Terrorists and other criminals may seek employment to gain access to information resources, to fund their activities or to provide legitimate activities to conceal their objectives. It is important to institute a strong personnel access security program.

Software Security

Fast-paced software releases virtually guarantee that software will be rushed to market before it has passed rigorous security scrutiny. In a hasty effort to meet deadlines and add enhanced, up-to-date features to their software, manufacturers often do not thoroughly security-test their products prior to their release. It is unconscionable to think that software manufacturers release products without thoroughly security-testing them, yet this is the reality of a competitive software marketplace. These manufacturers do this in hopes that security bugs can be fixed later by software patches. Therefore, a major portion of a security manager's job is to ensure that software being placed on the network has been designed with security in mind and thoroughly reviewed and tested prior to being placed on the network.

Assuming you have accomplished the aforementioned steps, your job of maintaining the software is not done. In fact, it is just beginning. Significant time may have passed since the time of the software's release from the developer; exposing previously unknown security flaws and weaknesses. It is for this reason that continued monitoring of all system software releases and updates is crucial to system security. Administrators should ensure system patches and updates are conducted as soon as possible.

Patches and fixes must be applied as soon as they become available, with one caveat; test software fixes on your test systems prior to administering them on your live network. This provides extra assurance that fixes do not cause damage to live network systems. Moreover, software fixes may have a ripple effect . . . causing unintended problems with other software on the workstations/servers. It is much easier to address these unintended consequences in a localized setting, such as a five-workstation test lab rather than on a live network of 7,000 workstations after the software patch has already been applied.

Identification and Authentication Methods

When a user provides identifying credentials, the authentication subsystem of the information system uses the credentials to authenticate the user to the system. In modern systems, passwords are checked against a stored, encrypted password file.

Though strong passwords are a deterrent against password guessing and cracking, they have a correlative side effect; as the requirements for the passwords elevate, users become more likely to write passwords down and store them where unauthorized users may access them. The methods most useful in preventing these insecure practices are user education and random sweeps through organizational areas. Users should be required to adhere to organizational policies regarding the safeguarding of passwords and must affirm their understanding of the policy.

User and administrator accounts that allow root access are particularly dangerous. If an attacker gains access to these accounts, they can read and delete files, and install 'backdoor' software to allow re-entry at a later time.[302] User education is a must to prevent reckless

---

[302] Panko, R.R. (2003). *Corporate Computer and Network Security*. Upper Saddle River, NJ: Prentice Hall Publishing.

handling of passwords and access cards and to prevent other hazards such as shoulder surfing, piggybacking and social engineering attacks. Dual user authentication methods are also a must.

Passwords were once the bastion upon which all system authentication routines were based, though recently taking an increasing role in two-factor authentication schemes (passwords and key cards, for example). The decreasing reliance upon password authentication is due to their dependence upon users to maintain password control. Passwords have often been compromised via social engineering attacks, weak password usage, leaving passwords unsecured, sharing passwords with unauthorized users or by other means. Strong user authentication schemes must rely on two-factor methods to establish that identity of the user.

Systems may use identifying credentials other than passwords to authenticate a user, such as crypto cards and keys, and biometrics. When used in conjunction with strong passwords, these methods enable an extra layer of protection. Most government systems now require all user systems to use a combination of two or more identification factors . . . typically a password and crypto card combination.

Biometric identification methods (e.g., fingerprint scanners, facial recognition, etc.) are relatively new, are expensive to implement, and require significant maintenance by qualified personnel. The slow acceptance of biometric authentication methods as a viable form of identification is due to their acceptance rate (the rate at which they deny access to authorized users and allow access for unauthorized users). As biometric technology becomes more reliable, easier-to-implement, and less expensive, it becomes more attractive to organizations. It remains to be seen whether biometrics becomes a mainstay for authentication.

Access Control

Let's discuss the internal operations that control access within a modern information system. After identifying and authenticating a user, access rights may be defined and given by the system utilizing a complex staging process. Within a system, access control rights may be governed by any one or a combination of mechanisms including automated terminal identification, and terminal logons to govern local and remote user logons. All accesses must be maintained in a system audit trail and continuously limited according to the system policies and the user's access rights.

Equally important is the governance of application access rights within a system. Applications should be implicitly denied access to system processes not expressly required for operation. During system and application development, objects should be designed to limit access and safeguard outputs. Most modern, proprietary operating systems provide these safeguards. Due to the adoption of a global architecture, called the Global Information Grid (GIG), government systems must be able to communicate and authenticate in a universally accepted way, which infers increasing reliance upon standardized identification, authentication, and access control methods.

Security 'Snooping' and Redundancy— The Often Overlooked Essentials

Proactive defense means occasionally stepping out from behind the desk and examining first-hand the systems under your charge. I call this 'snooping.' User diagrams tell only a portion of the story. Diagrams likely won't show you the potential emissions security hazards of errant equipment placement nor will they show you an unauthorized connection or a list of passwords displayed prominently above a user's workstation.

While working for one organization, I once experienced a major failure that catastrophically disconnected a network of 3,000 workstations and servers for almost an entire day. The failure was caused by a technician who simply placed a notebook over the cooling vent of a main network switch weeks before the failure. A network diagram will not show you this kind of catastrophic failure waiting to happen.

It is best to imagine how network systems might catastrophically fail. Are there any network components that are absolutely essential? Do you have replacement systems ready to take the place of vital systems? Is the data protected by redundant/backups? These kinds of questions enable you to envision a worst-case scenario that could destroy or disable your network or crucial network components. Tornado, flood, fire, theft, terrorist attack, cyber-attack . . . when developing a system security plan, try to envision every conceivable scenario in which you could lose assets/capabilities and develop a security plan to either repair/

replace the damaged assets in order to restore the network as quickly as possible following the event.

Do you have redundant storage of software/hardware assets? Are those redundant assets and facilities sufficiently geographically separated from the 'live' system? Do you periodically inventory and test redundant/backup systems to ensure their readiness in the event you need them? The best way to protect a system is to use your imagination to envision various disastrous scenarios and pre-determine the solutions or fix-actions that will counter them.

Managing Backup Media

System backups are the fail-safe mechanism of information security. If all precautions fail to prevent destruction of information and data, they will allow for the full or partial restoration of the system. As such, organizations must implement a backup program containing incremental and full system backups as necessary according to system requirements and organizational security policies. Periodically during the operation of network systems, users commit errors that cause irreparable damage to database files and other files. Incremental and full backups allow for the restoration of such files by the IT staff.

It has been the case that backups are conducted on a regular basis per the information policy, only to be found to be useless when actually required for restoration of data. Therefore, as with any other emergency procedure, it is crucial that the restoration procedure be practiced, tested and verified. Nothing should be taken for granted.

Tips for Managing Network and Remote Access Security

Computer assets are put in-place to support organizational objectives and the network should always be managed to provide maximum network availability to support them. All processes must be compartmentalized and accounted for within the information security policy; leaving nothing to chance. Where possible, encryption should be used to protect the transfer and storage of data and information. Patches for operating systems and data should be meticulously maintained. As connected or extra-organizational networks entail added risks for organizational systems, network partners should receive the same scrutiny as inter-organizational assets.

Network boundary firewalls should be capable of stateful inspection of incoming and outgoing packets. If possible, networks should be designed to defend against incoming attacks, including Distributed Denial of Service (DDoS) attacks in which a series of remotely controlled systems are used to overwhelm the network and deny access for legitimate traffic. These types of attacks can be

> *A "stateful" firewall is a firewall that keeps track of the state of network connections traveling across it. The firewall is programmed to distinguish legitimate packets for different types of connections. Only packets matching a known connection state will be allowed by the firewall; others will be rejected.*

carried out for a variety of reasons including extortion, market competition, political sabotage, or even cyberterrorism. Whatever the case, networks should be designed and operated to prevent unnecessary internal and external connections where possible. This minimizes internal and external network risk and threats and improves the overall performance of the network by limiting cross-traffic.

Threat/Risk Assessment

It is the duty of a security manager to determine the source and nature of the threat/risk, the level of risk posed, the threat exposure level, the ways in which a threat/risk may come to fruition and how those threats and risks may be countered, mitigated, or prevented. It is not necessary, nor even likely that all risks/threats can be envisioned or countered. However, by determining and developing plans for the most probable disastrous events, you take a major step towards preparing the system for plausible threats.

Threat/Risk management is the fundamental beginning of an organizational information security program and establishes the context by which the program is designed and implemented. The organizational security policy is derived from and continually refined by the results of the risk management process. Risk management for government organizations is principally not much different from that of civilian counterparts. However, government organizations must often wrangle with more complex regulatory requirements, intermingled with multiple government and non-government organizations and international entities.

Government organizations must often deal with the often antiquarian concepts of operation that characterize governments worldwide. The likelihood of exploitation is derived from the relative value of the information or information asset targeted by the attacker. Usually, risk exposure is calculated to reveal the cost effectiveness of the organization's risk management measures.[303] In regards to national secrets, the relative value must necessarily be assumed to be high; therefore, the impact of the potential risk is also surmised to be high.

Network Security Essentials

There is a pervasive and antiquated thought process propagated by some information security managers that we must defend by creating an impenetrable boundary to keep attackers out. This has often resulted in networks becoming constriction points and providing a false sense of security. What is the point in defending a network that becomes nearly unusable to the network users due to the course of our actions? Therefore, in developing a defensive strategy against cyber-attacks, we must keep in mind that defensive measures must always be balanced with system usability.

As many cyber-criminals have found it increasingly difficult to successfully attack well-defended networks, they have turned to more effective methods of gaining entry, including the use of social engineering attacks and virus software. System users are often the most overlooked "weakness" in an otherwise good security program. User training must be provided and must be mandatory to gain network access. User policies must be provided as a matter of course and enforced.

Many security managers face a constant stream of users attempting to increase the functional capacity of the network; either by official channels or by circumventing established security measures. As a new information security manager years ago in one organization, I discovered an entire wireless network consisting of ten personal computers that had never passed security scrutiny. These systems had been undiscovered and unknown to the previous organizational network security manager for over four years and had somehow been allowed to attach to the network.

---

[303] Pfleeger, C., & Pfleeger, S. (2003). *Security in Computing* (Third Edition). Upper Saddle River, NJ: Prentice Hall.

Information security involves tradeoffs. In many organizations, money is often devoted to security only after dire warnings . . . and begging and pleading by the information security staff. Warnings of impending doom are a good way to obtain funds, but will only get you so far before the spigot is turned off and your warnings begin to be discounted or ignored. Therefore, being truthful about threats and risks is the best policy for achieving long-term security program health.

It is crucial to evaluate and assess periodically the source, motive, likelihood, and potential severity of threats and to convey that information to upper-level management continually. Managers often pay only lip-service to information security . . . until a breach or damage occurs.

Security comes at the expense of other network necessities and niceties and is often lumped-in with other network items in the organizational requirements process for budgeting and forecasting purposes. This is a mistake. These are not 'convenience items' and 'nice-to-haves' as often are other network requirements. Security should always be a separate line item in the organizational budget along with forecasted and planned security enhancements.

Network Defensive Techniques

Firewalls, intrusion detection systems (IDSs), and the like must be replaced on a continuing and regular basis, regardless of whether operating system updates and software patches are being conducted. As with most network appliances there is only so much time before the operational capacity of these devices begins to degrade network efficiency and capacity. An old appliance become a network bottleneck, a potential point of failure and possibly risky to operate as technology passes them by.

Network protective measures should operate on the basic principle of 'Defense in Depth.' Defense in Depth simply means that no network protective measure stands alone. Redundant protection assures that at any entry point into the system an attacker will encounter at least two mechanisms designed to defeat an attack. For example, a network may have a firewall set to filter potential malicious content followed by similar filter residing on the organizational mail server. Redundant protection provides extra assurance that attacks are countered and that a mis-configuration of any one protective measure will not lead to a system compromise.

Another component of Defense in Depth to consider is that redundant protective systems should be dissimilar. For example, a network may contain CISCO firewalls on the perimeter and CheckPoint firewalls on the interior of the network. This ensures that the defeat of one security component due to an inherent flaw or programming error will not allow an attacker to bypass all network security measures as a result of the same flaw or exploit. However, this does entail additional work and training for network and information security personnel.

Finally, Defense in Depth should allow for internal segmentation of the network via security devices such as firewalls. An intruder that has defeated a perimeter network security system should never be allowed to roam freely about the internal network as a result. So to, segmentation decreases the ability of internal users to access systems and programs not under their purview.

Segmentation has many other uses. Segmentation on internal network firewalls can allow certain types of traffic to pass while blocking others, or may also allow only select systems to have access to the protected assets. For example, a computer on the operations floor may be allowed to print out documents on a printer within the accounting department while all other traffic is denied.

User Security

The weakest point in any corporate network is often its users. Too many security professionals rely upon technological solutions as a means to secure a network; often overlooking that users are often the origin of significant security failures. Technology cannot prevent a social engineering attack (discussed later) in which a legitimate user is tricked by an attacker posing as a legitimate network user or authority figure. There are a myriad of ways in which users may knowingly or unknowingly compromise an otherwise great information security program.

Human nature is also a persistent threat to information security. Typically, cyber-attackers and even authorized users feel it is their right to access and use organizational and even other user's personal assets for their own benefit.[304] Cyberterrorists likely feel even less respectful of the rights of ownership.

---

[304] Janczewski, L., & Colarik, A. (2005). *Managerial Guide for Handling Cyberterrorism and Information Warfare*. Hershey, PA: Idea Group Publishing.

Information Security Awareness in the Age of Terrorism

Though terrorism has often been mistakenly assigned a random nature, in reality, terror events are planned events that often take advantage of gaps in protection. By reviewing the underlying motives of terrorism and connected causal and preventive factors, organizations can take steps to minimize risk to information assets.

The difference between terrorism and other crimes is its underlying motivation. Terrorism is committed to instill fear, which drives the party being terrorized to react in a way that is considered to be favorable or positive by the terrorist or terror organization. Fear is usually not the final objective of the terror act; it is a means to achieve objectives.

Information Assets as a Target of Terrorism

Religious beliefs and associated philosophies, deep rooted hatred for the United States and its allies and the resentment associated with being a relatively wealthy country may all be primary reasons international terrorists use to motivate themselves to commit acts of terror. Homegrown terrorists may also commit such acts for similar or more localized reasons. Whatever the reason, information technology assets may be thought of as a primary target of interest to terrorists and terror organizations wishing to create significant economic and psychological damage to information oriented societies such as the United States.

Though information assets are obviously advantageous to any technology oriented nation, unprotected information infrastructures offer an inviting target for ambitious terrorists and can quickly become an Achilles heel to the societies that depend so heavily upon them. Network systems control a large share of other assets, such as telephone systems and subway trains. These have often been overlooked in the traditional view of information infrastructure protection.

Terror attacks can be directed upon IT assets, focused upon infrastructure items controlled by IT assets, or terrorists may even use these assets to further their agendas: Though information technology assets provide great economic advantages, their accessibility also provides advantages to terror organizations, who may use them to coordinate, direct or even launch attacks.

Everything from the development of critical information technology assets to the way they are employed and operated is affected by their perception as a potential and exploitable target; particularly following the attacks on September 11, 2001. For this reason, IT system and facility design must often go through a substantial review to ensure security, survivability, and maintainability in the event of an attack.

Cyber security

In the past, 'cyber security' was a very nebulous term. Security programs never gathered real attention from most corporate managers until the organization had either become the victim of a malicious attack or a significant high-profile attack had sufficiently alarmed them to the organization's vulnerability. Information security budgets were often lacking and security training for organizational members was nearly non-existent. Today, most organizational managers realize the stakes and the risks, yet most information security programs are still given only tangential thought. Why is that?

To analyze this problem, it is important to remember that security awareness is a cyclical event for most organizational members. In the days and months following the September 11 attacks, cyber security awareness peaked, with some managers becoming thoroughly convinced that professional cadres of al-Qaeda super-techies were preparing to wage cyberwarfare against U.S. financial institutions and military networks. Time and again, top government agencies issued warnings of imminent attack. Major professional organizations like the SANS Institute and InfraGard followed through to ensure that security managers received the latest news. When the predicted attacks did not occur, indifference to subsequent warnings often set in. It's a classic case of "Chicken Little" syndrome. Did these organizations overshoot their goals, thereby creating a climate of over-hyped cyber security concern?

The clear answer is "No." These organizations were simply performing the roles that all information security managers and organizations are tasked to perform, i.e., to anticipate and predict the moves of a cyber-attacker. It is not necessary that the warnings be correct to be effective. Indeed, it is the job of cyber security managers to anticipate the moves of their opponent . . . even if the anticipated attack is never realized.

Three essential items that malicious attackers must possess prior to committing a malicious cyber-attack include method, motive, and opportunity.[305] The method of attack depends greatly upon the skill level of the attacker and the tools at their disposal. Opportunity is governed by the time and access level of the attack; as well as by the access level of the victim of the attack. Motives vary greatly and provide the intentions of the attacker. Do they want to embarrass the victimized company? Do they want to destroy a resource or capability? Are they seeking financial gain from the attack? Questions like these provide the basis for developing a security plan. When you can anticipate an attacker's methods, motives, and opportunities, you can begin to prepare an adequate defense.

Cyberterrorism

Terrorists differ from other attackers in that they may be driven by an ideology. They may also be domestic or international in origin. It is important to review the motives of terrorism when determining what type of threat terrorists pose to your network. For example, a terrorist may be driven to deface a government organizational website to post detrimental information, cause destruction, or disrupt web services rather than attack a web server to steal credit card information for personal gain. Though this is so, one must be aware that terrorists always seek new ways in which to finance their activities. Hence, credit card theft and other for-profit crimes cannot be ruled out as a motive for terrorist network attacks.

A wide range of Internet-born attacks can threaten information resources and provide unauthorized access, which lead to fraud, theft, web defacement attacks, or other malicious or damaging results. The advantages of cyber-attacks are numerous for a cyberterrorist. With an amazingly low probability of being caught and an even smaller chance of being prosecuted, a cyber criminal may damage or incapacitate systems, steal or damage large volumes of data, deface websites for the humiliation of the organization, and in so doing, make a profound statement about the inability of an organization to stop the attacker. The

---

[305] Pfleeger, C., & Pfleeger, S. (2003). *Security in Computing* (Third Edition). Upper Saddle River, NJ: Prentice Hall.

ability to "defeat" network protective measures and deface and destroy a government web site would provide an opportunity for a terrorist to demonstrate their technical savvy and embarrass a superpower that has a decided technological advantage.

Network attacks may originate from countries that do not allow criminal extradition or are unwilling to assist an investigation into a web-borne attack. Hence, terrorist from those countries may remain anonymous and immune to prosecution. The cost of the attack may be equivalent to the cost of a personal computer, a network connection, some software, or a few hours of time. The toll of such an attack may range anywhere from defacing a government website to something more dramatic such as crashing a major financial network, thereby causing billions of dollars in losses. It is easy to see why these attacks are an attractive option for even a marginally well-financed terrorist organization.

It is important to remember that the only discernable difference between a terrorist with great computer skills and any other information warrior may be ideology. Though many of these organizations operate on shoestring budgets and are not particularly technically proficient, no one should disregard the motivation of these organizations and the threat they pose on that basis alone. It is unwise to discount these information warriors as being simple cobblers, farmers, and wristwatch repairmen who happen to own a computer. In fact, many reputedly possess sophisticated technical skills and advanced degrees in computer engineering and related fields. In order to dissuade a terror attack against your high-profile network, it is wise to accept the likelihood that the attacker possesses the method, motive, and opportunity to do so . . . regardless of preconceived notions.

It is worthy to note that some terror organizations, such as Hezbollah and HAMAS, are exceptionally well-funded and possess an abundance of personnel resources. To date, there have been no major network attacks known to have been attributed to these powerful regional organizations. However, that is not to say that these organizations can be discounted. It is safe to assume that it is only because network-born attacks have not been the primary focus of these organizations that they have not often engaged in them. In fact, it is almost a certainty that these powerful organizations maintain a cadre of capable information warriors willing to participate in cyber attacks for their extremist ideology at a moment's notice.

Types of Cyber-Attack and Cyberterror Attack

Unlike other forms of attack, which attempt to disable or destroy information assets, Denial of Service (DoS) and Distributed Denial of Service (DDoS) attacks create havoc by tying up network system resources with requests for services. The idea behind DOS and DDOS attacks is simple, i.e., to force a target system to become overloaded with activities that reduce its capacity to process legitimate tasks.[306] The attacker uses a series of zombie machines running hostile software to launch the attack that is unbeknownst to the owners of the zombie machines. Because these machines may be controlled remotely or even set to begin the attack at a predetermined time, tracing the attack back to the attacker may be extremely difficult. Even if it can be traced, the trail may ultimately lead to machines that are open to the public, such as in an Internet café.

It is difficult to predict or defend against DoS attacks. The viability of networked systems often hinges upon their ability to accept and answer the service requests that are often the basis of the attack. However, some solutions have been developed including lightweight packet-filtering mechanisms that provide enhanced traffic analysis.[307]

Some other solutions depend on using several intelligent packet-filters and redundant connections. However, these solutions are often useless when faced with an onslaught of thousands of zombie systems. Therefore, the prevailing thought behind most viable DoS solutions depends upon intelligent, stateful packet inspection capabilities. The general rule is to cut off the source of the attack.[308] Developing a rapport and a plan of defense against these attacks in cooperation with Internet service providers is also important.

---

[306] Janczewski, L., & Colarik, A. (2005). *Managerial Guide for Handling Cyberterrorism and Information Warfare*. Hershey, PA: Idea Group Publishing.

[307] Badishi, G., Herzberg, A., & Keider, I. (2007). Keeping denial of service attackers in the dark. *IEEE Transactions on Dependable and Secure Computing, 4(3)*. Retrieved February 1, 2008, from ProQuest database.

[308]

Web Defacement Attacks

These types of attacks are often viewed as mere nuisances. However, they may have far-reaching implications for an organization if they damage or destroy an organization's reputation or embarrasses the web site's owner. In the case of politically oriented attackers, attacks may be predicated upon embarrassing a government, government entity, or organization the attacker opposes. This type of attack may also be used to redirect traffic to a phony website or be combined with a phishing attack to steal information from persons visiting the defaced website, thereby leading to unauthorized account access or identity theft.[309]

Defending against web defacement attacks can be extremely difficult due to the simplicity of webpage design and the plethora of available design software packages containing varying security provisions. The current, standardized key to protecting against these types of attacks is to cut off the attack and provide component redundancy.

Although expensive and difficult to maintain, normally, an offline redundant copy of the website server and software is the key to quick restoral in the case of web defacement attack. Also, using a honey-pot system (a system created on the network to purposefully attract attacks and divert attention away from valuable assets) may distract the attacker and divert the attack away from the intended target of the attack. This approach allows the administrator to disconnect the attack, analyze its components, and gather information regarding the attack source.

Domain Name Service Attacks

Domain Name Service (DNS) attacks target DNS servers that provide domain name translation. By disabling a series of DNS servers, the attacker renders useless the resolving capabilities of the service, thereby, effectively producing a DoS attack. The DNS network protocol includes no provisions for authentication and, therefore, remains relatively open to attack. The attacker essentially gains access by impersonating legitimate DNS clients or redirecting legitimate network traffic.

---

[309] Janczewski, L., & Colarik, A. (2005). *Managerial Guide for Handling Cyberterrorism and Information Warfare*. Hershey, PA: Idea Group Publishing.

The answer to DNS attacks rests upon the introduction of IPsec authentication routines between DNS clients and servers. Another solution is to overlay the DNS protocol with an authenticating routine of its own, such as with DNSSEC. However, this method is greatly dependent upon all clients and servers using the DNSSEC overlay. Security administrators must ensure that one of these methods of authentication is used, that system patches are applied and security reviews are conducted on a regular basis.

Routing Vulnerabilities

Routers are the building blocks of networks. Routing capabilities allow the addressing and forwarding of traffic by specifying the locations of nodes on the network. As most routing protocols were primarily designed with reliability, speed and efficiency in mind, many security loopholes have been exposed and taken advantage of by attackers, who may listen to network traffic, hijack network sessions, initiate DoS and DDoS attacks, or other forms of attack. All of these attacks can happen at the nodes or the routers controlling network information flow.[310] Attackers often masquerade or spoof legitimate traffic in order to conduct attacks.

Most routing vulnerability attacks are built upon falsified network credentials, so authentication is an important first step to be taken to prevent attacks of this nature. However, using the IPSec protocol in IP networks would not be viable due to the resulting overhead. Therefore, the use of Public Key Infrastructure (PKI) is probably the best option as it provides strong authentication with moderate overhead. However, PKI can be difficult to implement and maintain.

Identity Theft and Social Engineering Attacks

Identity theft crimes are growing at an alarming rate in comparison to other theft crimes in recent years. A March 2007 study from the Gartner Group found that, from mid-2005 to mid-2006, about 15 million Americans were victims of fraud resulting from identity theft. This was an increase of more than 50 percent from the estimated 9.9

---

[310] Ibid.

million victims in 2003.[311] This rate of growth is only matched by the exponential increase in the level of sophistication these attacks reach as the cyber criminals learn to apply newer and more effective techniques to steal identities using increasingly creative ways. The motives for identity thefts are wide-ranging; intended to steal funds or cause damage to persons or organizations.

The most crucial yet often minimized aspect of physical security is prevention of social engineering attacks to obtain access to premises through the subversion of social standards. Personnel should be trained to look for and report suspicious activity and to protect against social engineering attacks that could inadvertently provide access to terrorists despite sophisticated and expensive physical measures. These attacks could be caused by shoulder surfing pass codes, piggybacking into the facility, or by pretending to be someone they are not, such as maintenance personnel, in order to gain access to unauthorized areas. Social engineering attacks, like other attacks, tend to follow the phased approach of target surveillance, selecting an area and method of attack, and the attack itself.[312] Once unauthorized access is provided, the terrorists may commit their acts of sabotage, damage, or other harmful deeds.

Social engineering attacks, also known as 'human hacking'or 'spoofing,' enable an attacker to elicit information from unsuspecting users to gain either personal information or information that may lead to further system or account access. In contrast to cyber-attackers using highly sophisticated computer skills, attackers using social engineering techniques gain access to the desired resources via deceptive methods perpetrated against persons possessing valid system access.[313] Some computer criminals attempt to exploit software programming errors. Attackers using social engineering techniques use methods that are comparatively uncomplicated, taking advantage of human trust and

---

[311] Fonte, E. (2008). Who should pay the price for identity theft? *Computer and Internet Lawyer, 25*(2). Retrieved February 1, 2008, from ProQuest database.

[312] Janczewski, L., & Colarik, A. (2005). *Managerial Guide for Handling Cyberterrorism and Information Warfare*. Hershey, PA: Idea Group Publishing.

[313] Ibid.

cognitive bias.[314] Attackers using social engineering techniques succeed because the victim allows them to succeed. A comprehensive user training program will assist to counter this threat.

The weak link in these types of attacks is the victim's lack of awareness concerning the problem. Though periodic user education will help to some extent, some users will always be susceptible to deception under the right circumstances. The remedy to identity theft can readily be compared to the remedies to network attacks—user education and challenge authentication—to name two. Technical security measures that may assist to prevent these attacks include providing encryption, authentication, and identification procedures.

Conclusion

Within this chapter, I have discussed some basic tenets of information security, security management, cyberterrorism, cyber-attacks, network security, and user security. This is only a brief overview of some basic principles of information security, and I encourage you to develop a greater understanding of the principles of information security and how they relate to the security of the organization by reviewing additional, comprehensive information security resources for more thorough coverage of the subject.

---

[314] Thapar, A. (2008). Social engineering: An attack vector most intricate to tackle! Retrieved September 25, 2008, from http://www.infosecwriters. com/text_resources/pdf/Social_Engineering_AThapar.pdf.

Chapter 10
Integrated Efforts in Fighting Cyberterrorism

William Shervey, MS, MBA

National, State, and Local Participation

In this section of this chapter, I identify contemporary efforts by national, state, and local governments to combat cyberterrorism. Secondly, I describe affiliations among the Department of Homeland Security (DHS), state, and local initiatives to combat cyberterrorism. I conclude this section by proposing the urgent need for all governing and non-governing bodies to participate in mitigating the dangers of new-world battlefields.

Introduction. In terms of world history, the global public has seen and experienced many dramatic, deadly, and well-documented terror incidents, which give reason and impetus for concern. American history was reasonably pristine until recent years. The bombing of the World Trade Center in the early 1990s comes to mind. The events of 9/11 come to mind. Many of us remember precisely where we were, what we thought, and how we felt when these events were broadcast around the world in real time. Many people considered these events as testimony to the dangers aimed at the very heart, soul, and existence of our country. As time passes, people tend to forget the horrid realities of what occurred on those two days. Our memories become faint. Maybe it is fear, maybe people are simply too busy with daily living, or maybe people think it couldn't happen to them. All of these probably play a part. Regardless of the reason, the reality is that the potential exists for these types of events to recur again.

Beyond Terrorism. The general public gives less attention to another danger. That danger is cyberterrorism. Cyberterrorism differs from terrorism in that perpetrators use computers and connectivity to implement their misgivings. Global computer connectivity allows access to and provides cyberterrorists with various methods to exploit adversaries for their own selfish benefit. Cyberterrorists use computers to communicate. They use these computers to promote and market ideologies and philosophies. They even use computers to gain access to critical infrastructures, which entire societies depend on for daily survival.[315] The DHS suggests that as the homeland becomes more difficult to attack physically, terrorists will turn to cyber attacks. The vulnerabilities of critical infrastructures present a fearful aspect of contemporary computer connectivity misuse.[316]

National and State Efforts to Combat Cyberterrorism. Genuine concern began in the mid-1980s for vulnerabilities in computers and connectivity working in concert with important critical infrastructures. Cordesman indicates a growing enthusiasm during this time about cyberterrorism and the potential impacts to critical infrastructure and ultimately to national security. Between 1985 and 2000, a number of high-level U.S. policy decisions established special investigative committees to research cyberterrorism and to provide required leadership. A number of senate subcommittee reviews have yielded stark evidence of vulnerabilities and the need to shore up defenses. Albeit a slow process, these initiatives have prompted substantive action in terms of implementing best practices in information and critical infrastructure protection. Incidents such as the bombing of the World Trade Center and the 9/11 attack have kept this concern alive and have fostered the spirit of a continued need.[317]

---

[315] Cordesman, A., & Cordesman, J. (2002). *Cyber-Threats, Information Warfare, and Critical Infrastructure Protection*. Westport, CT: Praeger Publishers.

[316] Department of Homeland Security (2007). *U.S. Homeland Security (Government & Private) Market Outlook—2007-2-11*. Retrieved July 12, 2008, from *www.homelandsecurityresearch.net/2007/01/15/us-homeland-security-government-private-market-outlook-2007-2011/*.

[317] Cordesman, A., & Cordesman, J. (2002). *Cyber-Threats, Information Warfare, and Critical Infrastructure Protection*. Westport, CT: Praeger Publishers.

Efforts to combat cyberterrorism in the United States have increased significantly in recent years. Fishman's work justifies these efforts citing a substantial increase in cyber attacks across public and private sectors.[318] The DHS estimates it will spend $28.5 billion between 2007 and 2011 in the private sector alone. Their research concludes that sectors within the industry will grow between 60% and 400%. These figures indicate activity at a national level with dollars trickling primarily to first-level contractors and then to tag-along subcontractors. It's reasonable to expect expenditures such as these to bode well for working Americans in terms of professional employability.[319]

Local Municipalities and Cyberterrorism. One aspect of the trickle-down effect supports national economic health. However, discontinuity is exhibited between the national efforts of the DHS and their ability to work consistently with state and local municipalities. In other words, a lot of money is being spent across the board, but there is little leadership on the part of DHS in directing state and local communities about how to best direct the funds. According to Johnson, decisions about how to apply budgets towards cyberterrorism are left to the discretion of state and local organizations.[320] Burkhammer clearly points out that partnerships between the DHS and state/local governments are "rare." This is unfortunate because, in part, the Department of Homeland Security is charged with the protection of cyberspace through training

---

[318]   Fishman, R., Josephberg, K., Linn, J., Pollack, J., & Victoriano, J. (2002). Threat of international cyberterrorism on the rise. *Intellectual Property & Technology Law Journal, 14*(10), 23.

[319]   Department of Homeland Security (2007). *U.S. Homeland Security (Government & Private) Market Outlook—2007-2-11.* Retrieved July 12, 2008, from *www.homelandsecurityresearch.net/2007/01/15/us-homeland-security-government-private-market-outlook-2007-2011/.*

[320]   Johnson, J. A. (2005). *Community Preparedness and Response to Terrorism: The Role of Community Organizations and Business, II.* Westport, CT: Praeger Publishers. Retrieved July 12, 2008, from Praeger Security International Online database: *http://psi.praeger.com.proxy1.ncu.edu/doc. aspx?adv_search=1&term_0=cyberterrorism&index_0=words&op_1=A ND&term_1=&index_1=words&op_2=AND&term_2=&index_2=word s&op_3=AND&term_3=&index_3=words&s=r&freeform=&d=/books/ dps/2000c277/2000c277-p2000c2779970041001.xml&i=0.*

and leadership.[321] Burkhammer's work further indicates that distribution of funds across all sectors is inequitable. Additionally, dollars forwarded to agencies are often used for first-responder items and not necessarily used towards the betterment of infrastructure security. The excellence is that we are addressing these issues, and improvement will be evident in the longer term.[322]

Concluding Remarks. Over the past 20 years, good work has gone on in cyber security and in mitigating vulnerabilities in critical infrastructures. It took some catastrophic events to gain the attention of lawmakers and policy setters. The pendulum currently swings in the right direction for reliability and necessary security to be designed into future systems. Care must be taken to watch opportunists who take unfair advantage of vulnerabilities to support their hidden agendas. Support for strong systems should come from top-down. All governing and non-governing bodies, both private and public, must focus on the needs of protecting critical infrastructure systems. The bottom line is that we must learn from the mistakes of others.

Local Level Initiatives

In this section of this chapter, I identify and discuss the purpose of local-level cyberterrorism initiatives. Further, I address the pros and cons of local initiatives and how national initiatives work or do not work in concert with local activities. Additionally, I identify and describe a local-level initiative currently on-going in Miami, Florida. I indicate that contemporary computers and connectivity methodologies are rife with vulnerabilities and that national, state, and local entities must work in concert to sustain the critical infrastructures that we enjoy as a nation and society.

Introduction. Over the past 25 years, technology advances have accelerated exponentially. Once requiring hundreds of square feet of floor space, computers today fit in the palm of our hand. Free-market capitalism and competition fuel the demand for smaller, cheaper, and faster computer systems. Computer chip advances propel the communications

---

[321] Burkhammer, L. (2006). The virtual enemy. *The American City & County, 121*(3), 32.

[322] Ibid.

industry forward. Much of today's voice communication is wireless. Beekman states that the Information Age has replaced the Industrial Age. Convergence of computer and communication technologies continues today.[323]

Both competing organizations and private citizens are the fortunate recipients of converged technologies. Smaller, cheaper, and faster computers provide many benefits for all users. Olson indicates that businesses today operate at the "speed of light." Smart technology enables them to keep up with competition. Organizations realize tangible improvements in efficiency and effectiveness. The private user can access vast amounts of virtual information. It is difficult to quantify the benefits of computers and connectivity that are at the public's disposal. To gain a sense of how societies depend on computers and connectivity, one needs only to ask the question: "What if all computers disappeared today?"[324]

There is another side of this story. Beekman suggests dangers associated with society's attachment to modern computers. Advanced computers, communication systems, convergence, an almost infinite level of information resources, and all the benefits that result from these things are not immune to criminal elements.[325] Taylor reports that computer-related crimes have been increasing in recent years. Perpetrators in societies spend more time circumventing societal protocol than they do making an honest living. They would rather spend their time exploiting others than being model citizens adding value to the greater good. Computers and connectivity simply provide a newer, more contemporary venue for perpetrators.[326]

Society, private/public organizations, governments, and non-governing bodies see the need to safeguard against illegal infiltration

---

[323] Beekman, G., & Quinn, M. (2006). *Computer Confluence: Tomorrow's Technology and You*. Upper Saddle River, NJ: Prentice-Hall.

[324] Olson, D. (2004). *Information Systems Project Management*. New York, NY: McGraw-Hill Companies, Inc.

[325] Beekman, G., & Quinn, M. (2006). *Computer Confluence: Tomorrow's Technology and You*. Upper Saddle River, NJ: Prentice-Hall.

[326] Taylor, R., Caeti, T., Loper, D., Fritsch, E., & Liederbach, J. (2006). *Digital Crime and Digital Terrorism*. Upper Saddle River, NJ: Pearson Prentice Hall.

and use of information. This threat affects all areas of society. Perpetrators compromise computer, network, and information infrastructures of many national, state, and local organizations. Pilfered pieces of data may not be very significant. However, a take-down of critical infrastructure represents a huge risk to society. Complex partnerships exist among national, state, and local entities. These relationships may compromise tight security and facilitate catastrophic attacks.

The Need for Local-level Initiatives. Computer-controlled infrastructures are pervasive throughout the United States. Computers and connectivity control and provide critical infrastructures such as water supply, electricity, banking, and transportation. This reality extends well beyond large cities and into rural communities of America for two reasons:

1. Most citizens, regardless of where they live, rely on critical infrastructure services.
2. Computers and connectivity are not limited by geographic location.

Therefore, I conclude that the need to protect these critical infrastructures extends deep into the heartland.

Much of the effort that goes into protecting critical infrastructure is in risk assessment and management. Johnson says risk techniques are "timeless facets of human nature." He suggests that ancient civilizations saw value in assessing risk and attempting to forecast potential outcomes. Thus, we may mitigate outcomes by being properly prepared. [327]

At the national level, the United States delegates the duties of risk assessment and management to formal authorities such as the DHS. However, the challenge of protecting all critical infrastructures is

---

[327] Johnson, J. A. (2005). *Community Preparedness and Response to Terrorism: III Communication and the Media.* Westport, CT: Praeger Publishers. Retrieved July 12, 2008, from Praeger Security International Online database: *http://psi.praeger.com.proxy1.ncu.edu/doc.aspx?q=local+cyberterrorism&imageField.x=0&imageField.y=0&newsearch=1&c =Monograph&p=0&s=r&newindex=&orig_search=&adv_search=1& num=1&freeform=&op_0=&term_0=cyberterrorism&index_0=words& d=/books/dps/2000c278/2000c278-p2000c2789970125001.xml&i=3.*

formidable. Hence, the DHS expects state and local entities to assume a substantial portion of the challenge. This isn't surprising since computers and connectivity often link critical infrastructures (CIs) in one physical location to other CIs in distant locations. Computer connections must be hardened to allow only necessary and appropriate access.

State and local governments involved in anti-terrorist and cyberterrorist activity employ a large number of people. This collaboration covers vast geographic regions. However, Thacher says that a decentralized, fragmented government may breed those very characteristics within state and local organizations charged with fighting terrorism.[328] To mitigate fragmentation in protecting the homeland, DeGaspari says the DHS suggests a layered approach to security. This layered approach promotes collaboration among the Coast Guard, Border Patrol, and private/state/local agencies, which results in a cohesive effort. This task is called the Container Security Initiative (CSI).[329]

Collaborations, however, face many challenges. An ongoing challenge today is for all departments to communicate seamlessly with each other. The aftermath of hurricane Katrina in New Orleans highlighted this communications inadequacy.

A Miami Florida Initiative. A Miami, Florida, initiative appears to provide good information to the public. A group with a longer history than cybercrime performs the work. The group is called Citizens' Crime Watch of Miami-Dade (CCWMD). Over the past few years, the director of CCWMD, Carmen Caldwell, clearly understood the need to inform the public about cybercrime issues. Using an existing information dissemination forum, Ms. Caldwell implemented a streamlined effort that combats immoral and illegal computer crimes at the local Miami level.[330]

---

[328] Thacher, D. (2005). The local role in homeland security. *Law and Society Review, 39*(3), 635.

[329] Degaspari, J. (2005). Layered Security. *Mechanical Engineering, 127*(5), 34.

[330] Caldwell, C. (2008, May 11). As We Get Older, Crimes Against Us Get Bolder, Miami, Fl: Retrieved July 22, 2008 from *Newsbank Online* database: *http://infoweb.newsbank.com/iw-search/we/InfoWeb?p_product=NewsBank&p_theme=aggregated5&p_action=doc&p_docid=121476C71CAAC2A8&p_docnum=15&p_queryname=4.*

In a recent article, Ms. Caldwell warned of emerging cyber scams that steal Internal Revenue Service (IRS) stimulus checks from recipients. She recently addressed the need for baby boomers to be aware of cyber scams that specifically target them. She points out that this burgeoning population segment is reasonably affluent. Hence, cybercriminals modify and tailor their tactics to target this affluent group.[331] Additionally, Ms. Caldwell warned of an ongoing email scam where unsuspecting computer users receive official-looking email stating that a complaint has been filed against them with the FBI. The email is convincing enough to trick some users into clicking a link or opening an attachment. When users do this, a virus embeds in their computer, which then steals password information.[332],[333] For child protection, Ms. Caldwell has written a number of articles highlighting the need for parents to be aware of online pedophiles.[334]

Ms. Caldwell and the CCWMD perform commendable work. The fundamental cybercrime purpose of CCWMD is to disseminate relevant information. Sometimes a little bit of awareness goes a long way.

---

[331] Caldwell, C. (2008, May 4). Beware of Internet Identity Thieves, Scams, Miami, Fl: Retrieved July 22, 2008 from *Newsbank Online* database: *http://infoweb.newsbank.com/iw-search/we/InfoWeb?p_product=NewsBank&p_theme=aggregated5&p_action=doc&p_docid=1209320AF457FE70&p_docnum=16&p_queryname=4.*

[332] Caldwell, C. (2008, June 29). FBI Sends Out Warning About Email Scams, Miami, Fl: Retrieved July 22, 2008 from Newsbank Online database: *http://infoweb.newsbank.com/iw-search/we/InfoWeb?p_product=NewsBank&p_theme=aggregated5&p_action=doc&p_docid=121F46FE29291430&p_docnum=1&p_queryname=2.*

[333] Caldwell, C. (2008, Feb 10). FBI Sends Out Warning About Email Scams, Miami, Fl: Retrieved July 22, 2008 from *Newsbank Online* database: *http://infoweb.newsbank.com/iw-search/we/InfoWeb?p_product=NewsBank&p_theme=aggregated5&p_action=doc&p_docid=11EBA3DE5C761260&p_docnum=34&p_queryname=4.*

[334] Caldwell, C. (2008, June 22). How To Keep Kids Safe During Summer, Miami, Fl: Retrieved July 22, 2008, from *Newsbank Online* database: *http://infoweb.newsbank.com/iw-search/we/InfoWeb?p_product=NewsBank&p_theme=aggregated5&p_action=doc&p_docid=121962B25B7868E0&p_docnum=3&p_queryname=3.*

Concluding Remarks. Large-scope collaborations are relatively new in American history. It is difficult for national, state, and local organizations to work harmoniously together. The Internet presents opportunities for perpetrators to take advantage of others. Initially, the Internet wasn't designed with security in mind. A certain amount of "trust" resides with computer users. Integration of converged systems with critical infrastructure goes unabated with the guiding goal of efficiency. Careful, thoughtful, and substantive planning must be the priority as we move forward. All private and public (national, state, and local) entities must be part of a collaborative effort to ensure that we move in a direction that maintains our existence.

International Practices Making Sense at the National Level

In this section of this chapter, I discuss new relationships between computers and telecommunication equipment and what that means to critical infrastructures. I also introduce Internet use through several different applications. Furthermore, I discuss how we use the Internet, which is similarly used by cyber criminals, foreign nations, and the United States. Additionally, I show how the emergence of cybercrime forces nations to collaborate with each another to be effective against it. Finally, I identify vulnerabilities in critical infrastructure and the need for diligence and vigilance to maintain its security.

New Systems Relationship. Contemporary computers, information systems, and telecommunications systems are tightly integrated. According to Beekman, powerful, small, and readily available computers sell at prices well within the reach of the population. Additionally, to use them, user-friendly computer interfaces require users to possess no in-depth computer science knowledge. Hence, the layperson is able to use the Internet and powerful computer programs.[335]

Powerful computers change the scope, use, and control of critical infrastructures. To take advantage of computing power, many nations create networks of tightly integrated computers, telecommunications, and critical infrastructure operations. Historically, this relatively new

---

[335] Beekman, G., & Quinn, M. (2006). *Computer Confluence: Tomorrow's Technology and You.* Upper Saddle River, NJ: Prentice-Hall.

relationship makes sense in order to create efficient and effective systems. However, Cordesman indicates that we must address critical infrastructure vulnerability concerns. Nations following this integration methodology should be mindful of closely guarding their critical infrastructures.[336]

Internet Use Beyond Critical Infrastructure: From Good to Bad. Beekman tells us that computers, connectivity, telecommunications, and the Internet feature numerous applications. The Internet is an excellent tool for global communication. Nations, states, groups, factions, communities, and criminals possess the ability to promote their beliefs and ideologies around the world in quasi-real time.[337] Kruse indicates that the Internet can be a tool for spying. Attackers can perform viable, cost-effective, clandestine missions from any location with an Internet connection. They use the Internet to maliciously deface public websites of adversaries. The Internet provides criminals with much greater reach. Their reach is no longer limited to national and international borders. Cyberterrorists possess the ability to project power in their favor and to undermine the confidence of peoples and countries.[338]

The International Community and Nations. The Internet is a global infrastructure. Very few countries are immune to its reaches. Authoritarian countries greatly restrict release of new technologies to their people. Democratic countries open new technology to their people. Developed nations around the world use computers and technology to accomplish a wide variety of tasks. As a result, most countries face ongoing challenges of cybercrime vulnerability. In general, criminals transfer their ill deeds from the streets to the Internet.

Evidence suggests value in the international effort to establish laws governing the use of new technology. Broadhurst suggests success with what he calls mutual legal assistance (MLA). He describes how members of the international community work together to formulate

---

[336] Cordesman, A., & Cordesman, J. (2002). *Cyber-Threats, Information Warfare, and Critical Infrastructure Protection*. Westport, CT: Praeger Publishers.

[337] Beekman, G., & Quinn, M. (2006). *Computer Confluence: Tomorrow's Technology and You*. Upper Saddle River, NJ: Prentice-Hall.

[338] Kruse, W., & Heiser, J. (2002). *Computer Forensics: Incident Response Essentials*. Boston, MA: Addison-Wesley.

laws that govern global Internet use.[339] Alexander and Bullwinkel suggest that international cooperation is absolutely necessary for countering cybercrime.[340],[341] Palfrey indicates that the European Council recommends the Internet as a surveillance venue to counter both crime and terrorism.[342]

How Are Cyber Criminals Using the Internet? Criminals develop many ways to use the Internet to support their activities. The crime of theft is ageless. Despite all good uses of the Internet, it also serves as a haven for theft. Identity theft headlines no longer alarm us the way they once did. Once a thief steals an identity, that groundwork promotes committing additional theft. Taylor says that terrorists use the Internet to promote their beliefs and philosophies. They use the Internet as a communication tool to reach the masses. They also use the Internet to recruit people sympathetic to their cause. Information which previously took weeks to disseminate now travels across the globe in seconds to minutes.[343]

Al-Qaeda uses the Internet to its best advantage. Lynch says that, even before 9/11, Al-Qaeda used websites to propagate its beliefs and philosophies. Spreading its message in this fashion also helps their recruiting. The leaders of Al-Qaeda, Osama bin Laden and Ayman

---

[339] Broadhurst, R. (2006). Developments in the global law enforcement of cybercrime. *Policing, 29*(3), 408.

[340] Alexander, Y. (2008). *Evolution of U.S. Counterterrorism Policy.* Westport, CT: Praeger Publishers. Retrieved July 27, 2008, from *Praeger Security International Online* database: *http://psi.praeger.com.proxy1. ncu.edu/doc.aspx?newindex=1&q=international+cooperation&image Field.x=0&imageField.y=0&c=&d=/books/gpg/C34692/C34692-2051. xml&i=0.*

[341] Bullwinkel, J. (2005). International cooperation in combating cybercrime in Asia: Existing mechanisms and new approaches. In R. Broadhurst, & P. Grabosky (Eds.), Cybercrime: The challenge in Asia (pp. 269–302). Hong Kong: Hong Kong University Press.

[342] Palfrey, T. (2000). Surveillance as a response to crime in cyberspace. *Information & Communications Technology Law, 9*(3), 173.

[343] Taylor, R., Caeti, T., Loper, D., Fritsch, E., & Liederbach, J. (2006). *Digital Crime and Digital Terrorism.* Upper Saddle River, NJ: Pearson Prentice Hall.

al-Zawahiri, understood the Internet as an equalizing force against much stronger opponents. They clearly used the Internet as a tool to their advantage.[344]

How Might the United States Fair Better in its Fight with Cybercrime? International efforts should be more warmly embraced by the United States to improve cyber security. The United States needs to be forthright in its cooperation with other nations and needs to improve collaborative efforts around the globe. Additionally, the United States should adopt stronger policy in assisting the international community with complex and diverse legal aspects of governing Internet activity. Further emphasis needs to be placed on formulating appropriate legal response to undesirable cyber use. Additionally, the United States should embrace stronger surveillance measures to neutralize cybercriminals.

Concluding Remarks. Contemporary technology advances at an astounding rate. This advancement offers many new ways to implement public infrastructure services. One needs only to look back a couple of decades to see remarkable differences between old and new critical infrastructure systems. Unfortunately, inherent vulnerabilities in these systems provide opportunity for perpetrators to take advantage of victims. For the Internet to remain a global conduit, the United States must maintain a more global involvement in the policy equation. This requires international and national communities to embrace closely all aspects of cyber security.

---

[344]   Lynch, M. (2006). *Al-Qaeda's Media Strategy*, Retrieved July 31, 2008, from *The National Interest Online*: *http://www.nationalinterest.org/printerfriendly.aspx?id=11524*.

Chapter 11
Computer Security, Forensics, and Cyberterrorism

William Shervey, MS, MBA

Cyberterrorism, Digital Forensics, and Legal Challenges

The purpose of this section of this chapter is twofold: First, I discuss the difficulties in collecting computer crime evidence. More importantly, I cover how crime scene evidence must be treated throughout its lifecycle to ensure its admissibility in U.S. courts. Second, I present the numerous obstacles that the international community faces in counteracting cybercrime.

Computer Forensics. In recent years, crime scene forensics has captivated the general public. Kruse supports this notion by revealing the numerous television shows that depict crime and shows how investigators dissect minute details.[345] Why does the public gravitate to crime scene forensics? Kruse believes that it is based on fascination. Regardless of reason, Kruse suggests that spirited interest is good because it projects forensics into a limelight where it receives needed support.[346]

Within the body of crime scene forensics, we find a niche known as computer or digital forensics. Computer forensics is the "preservation, identification, extraction, documentation, and interpretation of computer data."[347] Taylor tells us that digital forensics relate to a number of civil court applications such as computer network analysis and software

---

[345] Kruse, W., & Heiser, J. (2002). *Computer Forensics: Incident Response Essentials*. Boston, MA: Addison-Wesley.

[346] Ibid.

[347] Ibid., p.2.

program analysis. However, he quickly points out that most recent proceedings deal with computer storage aspects of computer forensics called imaging.[348] We learn from Beekman that a *medium* is that part of a computer that stores information. A number of different types of storage media include hard drives, floppy disks, thumb drives, and compact disks (CDs).[349] Taylor reports these as static media since they store dormant information. Computer storage media are often the target of digital crime scene investigations.[350]

Difficulties in Using Digital Evidence in Court. Kruse explains that any successful investigation hinges on complete documentation. This is particularly true of computer storage forensics. As forensic experts testify in court, they must convince a presiding judge that the evidence is tamper free. Any indication that evidence has not been strictly safeguarded during the entire investigation lifecycle may prompt a dismissal of the case.[351] Kruse promotes computer forensics as a process which includes: (1) acquisition (2) authentication and, (3) analysis.[352]

Cybercrime must be treated on a case-by-case basis. However, Kruse indicates that many computer forensics investigations begin with acquiring source material from a suspect computer. If the target computer is in the field, a court magistrate may need to provide a warrant for investigators. In order to remain within 1st amendment rights, a magistrate might be inclined to issue a least-invasive warrant. Adding complexity to initial image acquisition begs the question: "Does the computer's random access memory (RAM) contain valuable information?" For an operating computer, RAM contains "live" information about what the computer is currently processing.[353] However,

---

[348] Taylor, R., Caeti, T., Loper, D., Fritsch, E., & Liederbach, J. (2006). *Digital Crime and Digital Terrorism*. Upper Saddle River, NJ: Pearson Prentice Hall.

[349] Beekman, G., & Quinn, M. (2006). *Computer Confluence: Tomorrow's Technology and You*. Upper Saddle River, NJ: Prentice-Hall.

[350] Taylor, R., Caeti, T., Loper, D., Fritsch, E., & Liederbach, J. (2006). *Digital Crime and Digital Terrorism*. Upper Saddle River, NJ: Pearson Prentice Hall.

[351] Kruse, W., & Heiser, J. (2002). *Computer Forensics: Incident Response Essentials*. Boston, MA: Addison-Wesley.

[352] Ibid.

[353] Ibid.

Beekman says, by nature of its design, RAM is volatile. When a computer is powered down, its contents are gone. At times, "live" information proves valuable to an investigation. Since RAM is volatile, an investigator may need to capture its information prior to turning off the computer.[354]

Digital forensic experts may be required in court to explain technicalities to non-technical people. To maintain image authenticity and reliability, a prosecutor normally wants an expert to demonstrate to the court that they took all possible actions during the evidence lifecycle.

Demonstrating authenticity shows the court that all evidence or derivations of evidence comes from an authenticated source. From a technical standpoint, investigators may use hashing software to authenticate evidentiary images. Kruse and Taylor describe the hashing process. An investigator first runs hashing software against a target computer's hard drive. As a result of running hashing software, a digital fingerprint in numeric form is provided, which is unique to the image of that hard drive. Next, using special copy software, the investigator meticulously makes an exact copy of the target hard drive and places it on a clean, secondary hard drive. The same hashing software is run on the secondary hard drive. If the image copy was successful, the hashing software should yield the identical numeric, digital fingerprint as did the target drive. Additional care must be taken during image copying not to "write" new information on the original target hard drive. In court, this might be construed as unintentional tampering but tampering none-the-less. An investigator may find it advantageous to run hashing software at various times throughout the lifecycle of the evidence. Once the investigator produces appropriate images, they are ready for analysis. Case circumstances typically dictate the required type of analysis. The case may require analysis ranging from child pornography to fraud. Many types of software analysis tools are available.[355], [356]

National and International Law Struggles to Remain Current. Schultheis portrays the late 20th century as computer convergence.

---

[354] Beekman, G., & Quinn, M. (2006). *Computer Confluence: Tomorrow's Technology and You.* Upper Saddle River, NJ: Prentice-Hall.

[355] Kruse, W., & Heiser, J. (2002). *Computer Forensics: Incident Response Essentials.* Boston, MA: Addison-Wesley.

[356] Taylor, R., Caeti, T., Loper, D., Fritsch, E., & Liederbach, J. (2006). *Digital Crime and Digital Terrorism.* Upper Saddle River, NJ: Pearson Prentice Hall.

Through advancing technology, we saw legacy communication systems converge with computer systems. Over time, converged systems become more robust as R&D produces faster, smaller, and lower-cost products.[357] Grabosky says that computer and communication convergence significantly impacts the way we live.[358] While many of these changes are positive, there is also a dark side. According to Grabosky, we see criminal activity, which was unforeseeable 20 years ago. He says that as computer products have become more user-friendly, a larger segment of society possesses the "capacity to inflict massive harm."[359]

Cordesman points out that ill-prepared international law deals with cybercrime. New technologies present opportunities for new crimes. Newness is important because it means that laws to deter new crimes may not yet exist.[360] Grabosky reports that most legal systems abide by the *nullum crimen sine lege* principle, which means that regardless of how bad our actions may be, we may not be prosecuted unless we violate a specific law. He exemplifies this concept explicitly by pointing to the I LOVE YOU virus. In this case, the perpetrator could not be prosecuted because there were no existing laws that addressed the release of malware. Laws are not created in real time. Instead, they culminate over the course of ongoing judicial proceedings. The speed at which technology has developed demands legislators work quicker to create legal deterrence.[361]

Another part of changing law encompasses how we collect and verify digital forensics. Grabosky supports the notion that prosecutors should involve themselves in digital forensics in the early stages of investigation. He points to prosecutorial awareness and understanding and how much and what type of evidence they must obtain for convictions. Knowledgeable

---

[357] Schultheis, R., & Sumner, M. (1998). *Management Information Systems.* Boston, MA: Irwin/McGraw-Hill.

[358] Grabosky, P. (2007). Requirements of prosecution services to deal with cybercrime. *Crime Law Soc Change, 47*(1), 201.

[359] Ibid.

[360] Cordesman, A., & Cordesman, J. (2002). *Cyber-Threats, Information Warfare, and Critical Infrastructure Protection.* Westport, CT: Praeger Publishers.

[361] Grabosky, P. (2007). Requirements of prosecution services to deal with cybercrime. *Crime Law Soc Change, 47*(1), 201.

prosecutors can foster effective investigations and protect the integrity of the investigative process.[362]

Compounding cybercrime and cyberterrorism is its international component. Global computer networks create a connectivity, which generally does not respect physical, sovereign borders. As a result, perpetrators have been empowered to commit crimes in any nation and from any nation. Issues quickly become muddied through the dictates of two or more sets of national laws, many of which are in disagreement. Issues of jurisdiction often create friction. As Grabosky points out, tracking these perpetrators can be difficult because, fundamentally, the Internet was not designed with security in mind, and tracking illicit activity is fragile at best.[363]

How are Local and International Authorities Responding to Cyberterrorism? Many developed and developing countries share concern over cybercrime which transcend borders. Grabosky cites collective interest in areas such as human rights, international cooperation, mutual legal assistance, jurisdiction issues, foreign prosecution, extradition, inappropriate charges, prosecution of juveniles, and what role the private sector might play in cybercrime.[364]

Some authors address the following cybercrime law policies:

- *Laws.* Enactment of substantive and procedural laws that cope with current and anticipated manifestations of cybercrime
- *Skills.* Development of forensic computing skills by law enforcement/investigative personnel and judicial officers
- *Harmonization.* Achievement of legal harmonization ideally at a global level
- *Cooperation.* Creation of mechanisms for operational cooperation between law enforcement agencies from different countries—24/7 points of contact for investigators and mechanisms for mutual assistance in general cyber-criminal matters.[365]

---

[362] Ibid.

[363] Ibid.

[364] Ibid.

[365] Bullwinkel, J. (2005). International cooperation in combating cybercrime in Asia: Existing mechanisms and new approaches. In R. Broadhurst & P. Grabosky (Eds.), *Cybercrime: The challenge in Asia* (pp. 269–302). Hong Kong: Hong Kong University Press.

Concluding Remarks. Cybercrime, cyberterrorism, and digital forensics are all relatively new phenomena. As such, local and international governments struggle to remain effective in proposing and passing digital crime legislation. Current consensus fosters the spirit of strict processes in the collection and analysis of digital evidence. At local levels, nations strive to bolster legal prosecution of many new types of digital crime. At the international level, nations are realizing the need to cooperate with other nations to mitigate transnational violations. As technology continues to advance, all peoples of the world will be at risk and will need to be involved in mitigating cybercrimes.

Cyberterrorism: Much Done, Much Yet to Do

Beekman tells us that the technology landscape is dynamic and continually changes over time. This phenomenon has never been more apparent than in recent history. Innovation in new technology means advances in medicine, business, transportation, and critical infrastructure.[366] Janczewski suggests that the power of contemporary technology is a tool for communication. That communication rallies behind philosophical and societal beliefs. Positive outcomes of technology advancement do not stand alone.[367] According to Taylor, perpetrators use computers and the Internet to violate the free rights of others through psychological and physical cyberterrorism and cyberwarfare.[368]

In this section of this chapter, I discuss the progress made in countering cybercrime and cyberterrorism. To continue that progress, the United States must remain diligent in five central areas as it moves forward in the rapidly changing dynamics of new world technology.

Introduction. While we have expended much effort towards countering cyberterrorism and cyberwarfare, much work still needs

---

[366] Beekman, G., & Quinn, M. (2006). *Computer Confluence: Tomorrow's Technology and You.* Upper Saddle River, NJ: Prentice-Hall.

[367] Janczewski, L (2005). *Managerial Guide for Handling Cyberterrorism and Information Warfare.* Hershey, PA: Idea Group Publishing.

[368] Taylor, R., Caeti, T., Loper, D., Fritsch, E., & Liederbach, J. (2006). *Digital Crime and Digital Terrorism.* Upper Saddle River, NJ: Pearson Prentice Hall.

to be done. Cordesman indicates that the United States struggles with defining *cyberterrorism*, *cyberwarfare*, and *critical infrastructure*. However, we have made good progress in assessing the need for and designing policy/procedure.[369] Peltier advocates the development of policies and procedures as a project. Much work has been focused on developing programs and defining needs. The United States continues to strengthen cyberwarfare capabilities. We have made much progress in determining what working relationships should exist among national, state, and local affiliations.[370] Alexander indicates that there has also been genuine acknowledgement for the need to cooperate and collaborate on an international scale. The United States also recognizes and works towards identifying the differences between private and public sector critical infrastructures and understands that we must require different incentives to improve these areas.[371] Cordesman further suggests that a considerable gap exists between how civilian and military sectors view cyberterrorism, cyberwarfare, and critical infrastructure protection. How the United States should proceed is often contentious between these two sectors. From a project management standpoint, public and private sector organizations struggle to address the many organizational issues surrounding cyberterrorism, cyberwarfare, and critical infrastructure protection.[372]

Cordesman explores various aspects of cybercrime, cyberterrorism, and critical infrastructure protection. These topics are both complex and

---

[369] Cordesman, A., & Cordesman, J. (2002). *Cyber-Threats, Information Warfare, and Critical Infrastructure Protection*. Westport, CT: Praeger Publishers.

[370] Peltier, T. (2004). *Information Security Policy and Procedures*. Boca Raton, Fl: Auerbach.

[371] Alexander, Y. (2008). *Evolution of U.S. Counterterrorism Policy*. Westport, CT: Praeger Publishers. Retrieved July 27, 2008, from *Praeger Security International Online* database: *http://psi.praeger.com.proxy1. ncu.edu/doc.aspx?newindex=1&q=international+cooperation&image Field.x=0&imageField.y=0&c=&d=/books/gpg/C34692/C34692-2051. xml&i=0.*

[372] Cordesman, A., & Cordesman, J. (2002). *Cyber-Threats, Information Warfare, and Critical Infrastructure Protection*. Westport, CT: Praeger Publishers.

diverse, and they encompass concepts ranging from highly technical in nature to pure project management.[373] Cordesman's final chapter lists 30 Department of Homeland Security recommendations. A number of the recommendations overlap. The following paragraphs provide a comprehensive but concise aggregation of those 30 recommendations. They have been synthesized into five most important areas.[374]

Definitions, Vulnerability Identification, and Priorities. The first area of aggregation precisely defines terminology such as *cybercrime*, *cyberterror*, and *cyberwarfare*. It may be useful to consider the root words of these terms in order to define them. Crime, terror, and warfare each have their own meanings and prefixing each with "cyber" decidedly relates them to computers and connectivity. In any event, we understand the need for equitable definitions across all sectors. For further understanding, these definitions need to be distinguishable from each another.

Work must be performed to identify real critical infrastructure vulnerabilities. When all vulnerabilities are identified, they must be prioritized from most to least significant impact. Then, beginning with the most important vulnerabilities, a series of assessments should be performed. Real critical infrastructure vulnerabilities are assessed with regard to deterrence, defenses, and attack response. Additional work must focus on infrastructure isolation, alternatives, and remediation. If particular vulnerabilities are not defensible by the U.S. government, then special attention must be placed on delegating effort to the private sector.

Groups and Their Obligations. Cordesman reports that protecting critical infrastructure remains a large and daunting task. Many infrastructures in the United States and globally are considered critical. The tremendous geographic dispersion and number of critical infrastructures requiring protection cannot rest solely on the federal government. Responsibility and delegation of duties must be shared. Responsibility begins at the federal level. A clear understanding must exist among national, state, and local officials. Boundaries must be established and responsibilities allocated. While there may be overlap, groups that need to step up and assume responsibility are divided into

---

[373] Ibid.

[374] Ibid.

five categories: (1) national, (2) state, (3) local, (4) public sector, and (5) private sector. Each group must possess a clear scope of obligations and responsibilities. The effort required to combat cybercrime/cyberterrorism and to protect the homeland critical infrastructure is too vast to assign to a single group.[375]

Talent Pools and Genuine Support. Profit generally drives private sector endeavors. Therefore, private sector organizations usually attract top performers with attractive financial incentives. Alternatively, public sector organizations do not possess this leverage. Public and private sector employees generally are not compensated at the same level. Thus, governmental organizations maintain a significant challenge in attracting top performers. This does not imply that good people do not work for the government. Rather, government cannot readily attract top performers across the board. Also, training and retention of these employees represent a systemic challenge.

Project and Program Management. Width and breadth of today's technology mandate the need to create a single or central point of authority for cyber-intelligence, cyberwarfare, and cyber-defense. These items represent, in part, focal areas for the DHS. Funding issues for this department are critical and financial allocations must be studied carefully. International and national efforts must be considered. The DHS must clearly address the issues of being prepared for cyberwarfare with extremists, terrorists, and other governments.

Given the size and scope of this work, strong leadership is essential in program and project management. Olson says that good program managers work through others to influence and achieve buy-in from stakeholders. From the public sector viewpoint, the United States needs to recruit the best program managers with large-project experience. People with this type of experience must be lured away from the private sector where their compensation is much higher. Regardless, this must be a serious consideration of all public sector service providers.[376]

Legal Issues. Equal time must be spent analyzing the legal aspects of cybercrime, cyberterrorism, and cyberwarfare. Much of this territory is uncharted, and laws for incarcerating cyber-criminals simply do not

---

[375] Ibid.

[376] Olson, D. (2004). *Information Systems Project Management*. New York, NY: McGraw-Hill Companies, Inc.

exist. Insufficient case law exists to litigate accused cybercriminals. Additionally, going forward requires a comprehensive review of critical-infrastructure liabilities and criminal law.

If attacked, the United States must maintain the legitimate right to defend its homeland critical infrastructure. Similarly, the United States must possess the legal right to attack offensively if necessary to circumvent future attacks on critical infrastructure. Without question, international law needs to evolve in scope to include all aspects of cybercrime, cyberterrorism, and critical infrastructure protection.

Concluding Remarks. Cybercrime, cyberterrorism, and critical infrastructure protection present many challenges. Over the past 25 years, much work has been performed in these areas. However, moving forward, work needs to be done. We must establish and understand common definitions at all levels. We must identify and prioritize (by importance) real vulnerabilities. We must establish clear groups of authority, and these groups must recognize and perform their responsibilities. We must establish strong talent pools in the public sector venue. We must provide these employees with genuine support. We must strengthen program and project management with qualified individuals who can cope with diversity. International collaboration must be part of the culture. Finally, a strong effort must be made to handle all legal issues related to cybercrime, cyberterrorism, and critical infrastructure protection.

Chapter 12
Protecting Our Critical Infrastructure

William Shervey, MS, MBA

The Emerging Role of Computer Security in Critical Infrastructure Protection

T he purpose of this section of the chapter is to define the words *infrastructure* and *critical infrastructure*. I also describe how critical infrastructures have evolved to rely on computers and the Internet. Both public and private sector organizations rely heavily on current technology in order to operate and maintain critical infrastructures. Additionally, in this section, I discuss organizations' obligatory and legal responsibilities with respect to protecting infrastructures. I conclude with a discussion of the difficulties implementing organizational forensic programs.

Introduction. *Infrastructure* is defined by *Merriam Webster Online* as "the underlying foundation or basic framework."[377] When the word *critical* is combined with *infrastructure*, a different meaning is established. *Wikipedia* describes *critical infrastructure* as assets required by societies and economies.[378] The following infrastructures fulfill the needs of large populations:

---

[377] "Infrastructure." *Merriam-Webster Online Dictionary*. 2008. *Merriam-Webster Online*. Retrieved May 14, 2008, from *http://www. merriam-webster.com/dictionary/infrastructure*.

[378] "Critical infrastructure." *Wikipedia, The Free Encyclopedia*. 1 Apr 2008, 22:30 UTC. Wikimedia Foundation, Inc. Retrieved May 14, 2008, from *http://en.wikipedia.org/w/index.php?title=Critical_infrastructure&oldid =202658075*.

- Electrical power production and distribution
- Communication
- Water cleanliness and distribution
- Food production and distribution
- Medical assistance
- Transportation (people and product movers)
- Military and police
- Financial services

Computer and Connectivity Reliance. Many of these critical infrastructures are tenured by time. More recently, however, critical infrastructures have evolved into mechanisms and processes which rely directly on computer systems and connectivity to sustain them and the services they provide. According to Cordesman, these types of changes to infrastructure often occur at a rate which is too fast for governments to keep pace.[379] Smarr refers to computers and connectivity as a global fabric and predicts integration as seamless and invisible in the future. The current course of computer and connectivity integration into infrastructures appears to be irreversible.[380]

In support of contemporary connectivity, research by Weinstein indicates that the National Science Foundation recently awarded $35 million to design TeraGrid. The TeraGrid concept is distributed computing and sometimes is referred to as grid computing. It dispels the old notion of using a single mainframe for computing. Rather, grid computing creates a network of computers that share workloads to improve performance. Another grid computing network being constructed is called National Lambda Rail (NLR). Grid computing efforts are being funded with millions of dollars directed towards developing future systems. The immediate effect of this funding boosts economies as the money trickles down to subcontractors. Longer term goals include supporting infrastructure effectiveness and realizing efficiencies. As time passes,

---

[379] Cordesman, A., & Cordesman, J. (2002). *Cyber-Threats, Information Warfare, and Critical Infrastructure Protection.* Westport, CT: Praeger Publishers.

[380] Smarr, L. (1999). Digital fabric. *Research & Development, 41*(7), 50.

critical infrastructures become more integrated and reliant upon computers and connectivity.[381]

Public and Private Sector Involved. In the United States, some infrastructures result from private enterprise and capitalistic endeavors. Others are sponsored by and championed through government workforces that are financed by public taxes. Still, others are distinguished as hybrids because they compete in the open market and receive government subsidies. In general, electrical, communications, and food production and distribution comprise examples of private infrastructures. Public infrastructure examples include water treatment and distribution, militia, and police departments. Examples of hybrid infrastructures include education, health, and financial. In terms of education hybrids, Caucutt suggests that additional subsidies would provide a positive effect on inequalities, welfare, and efficiencies.[382]

Generally, the U.S. government supports publicly funded militias and police infrastructures. These infrastructures are required to protect against outsiders and lawbreaking insiders who disrespect generally accepted societal behaviors. Entrepreneurs compete in the open market to provide electricity, communications, and foodstuff to society. Volokh argues that privatization of certain sectors may invoke a more selfish attitude among product and service providers. An in-depth examination of why infrastructures are public, private, or hybrid is beyond the scope of this chapter. However, it is important to understand that they do exist, are fundamentally grounded in computers and connectivity, and that existing vulnerabilities require attention.[383]

Protecting Critical Infrastructure. Not surprisingly, critical infrastructure assets are not always resilient to external influence and potential harm. Harm to infrastructure may be caused by natural

---

[381] Weinstein, L., & Clower, T. (2004). Grid computing and an optical fiber network: How they can bolster the Texas economy. Retrieved from *www.thecb.state.tx.us/reports/pdf/0723.pdf.*

[382] Caucutt, E., & Kumar, K. (2003). Higher education subsidies and heterogeneity: A dynamic analysis. *Journal of Economic Dynamics & Control, 27*(8), 1459.

[383] Volokh, A. (2008). Privatization and the law and economics of political advocacy. *Stanford Law Review, 60*(4), 1197.

disasters. Destructive weather damages infrastructure. Fire damages infrastructure. Other dangers are manmade such as maliciousness of those desiring to harm infrastructures. Reasons and results of manmade maliciousness vary. However, infrastructure sponsors must protect their infrastructures to ensure continued operations.

Organizations spend much time on business continuity planning (BCP). Botha supports this effort by stating that contemporary organizations must operate continuously to survive, and that BCP is one way to make that happen. Good BCP establishes a plan that the organization implements when infrastructure capabilities are crippled or impaired. In addition to providing procedures to return infrastructures to normal operations, BCP should address lessons learned from negative events.[384]

Forensics and Critical Infrastructure: The Difficulties. Advancing computer and connectivity capabilities provide organizations with opportunities to increase performance, enhance effectiveness, and improve efficiencies. Unfortunately, innovation is a double-edged sword. Some use innovation to take advantage of others. Reasons vary widely as to why they do this. Some seek the thrill of infiltrating a hardened infrastructure. Others disrupt and bring down infrastructures because of their ideologies and beliefs. Successful maliciousness that renders infrastructures inoperable places society at risk.

Physical security protects infrastructures and the computers that enable their operations at acceptable levels. However, nobody owns the Internet, which is available for everyone's use. Hence, perpetrators can access the same systems as infrastructure supporters. Internet connectivity, when neither hardened nor protected properly, provides backdoors for malicious activity. This vulnerability may remain unresolved in the foreseeable future.

Conclusion. In the past, infrastructures were far less sophisticated than they are today. With passage of time, populations have grown throughout the world. Concurrent with population growth, a rising need exists to provide additional, required infrastructures to support society. Meanwhile, computers grow more powerful and are continually integrated with infrastructures for improvement. In the United States,

---

[384] Botha, J., & Von Solms, R. (2004). A cyclic approach to business continuity planning. *Information management & computer security, 12*(4), 328.

both private and public sector organizations heavily rely on computers, information systems, and Internet connectivity as primary components of critical information systems. As these infrastructures grow in complexity, they are scrutinized by those who desire to undermine them. As a result, organizations must work hard to remain ahead of maliciousness. Hence, as infrastructures become more complex, so do efforts to guard and keep them operating safely.

U.S. Legal Requirements and the Difficulties in Forensic Applications

In this chapter's section, I cover various aspects of cyber security and how they relate to critical U.S. infrastructure. Additionally, I provide historical insights to the legal implications of computer and cyber security in the United States. Furthermore, I highlight difficulties with applying forensics in a rapidly changing technological environment. Finally, I propose that, despite rapidly changing technical environments, a real need exists for protecting contemporary critical infrastructures.

Introduction. Critical infrastructure in the United States plays a significant role in our daily lives. How about on-demand electricity? How about clean water? Where do we keep our money, and how do we conduct financial transactions with others? How do we communicate over long distances? How extensively do we rely on work and home computers and their connectivity to the Internet? How much do we rely on transportation systems? In contemplating these questions, we understand that existing without one or more of these infrastructures would highly inconvenience us. Wagner says that contemporary critical infrastructures are so engrained in our daily lives that damage to any of them would instantaneously impact us.[385]

Changing Times: Changing Policy. As computers evolve, they become more powerful and less costly. Public and private sector organizations legitimatize investment in computing resources by bringing attention to the potential efficiencies and effectiveness that may result. Few aspects of society escape computer integration. Consequently, computing resources today play a significant role in U.S. critical infrastructures.

---

[385] Wagner, C. (2007). Countering cyber attacks. *The Futurist, 41*(3), 16.

The U.S. government is resolutely concerned with protecting and sustaining critical infrastructures. Cordesman indicates that in the mid-1980s, the federal government acknowledged the presence and threat of cybercrime. In so doing, the government passed the first substantive legislation in the United States that specifically addressed information cyber security. Passed in 1987, the Computer Security Act established procedures and guidelines for federal computer systems. The act also prompted government organizations to adopt security guidelines developed by the National Security Agency (NSA).[386]

According to Cordesman, the government's Computer Security Act was successful only to a certain degree. Below the federal level, state and local governments were not fully participating in the Computer Security Act.[387] Cordesman indicates that this lack of action gave rise to a renewed interest in computer security. Therefore, in 1995 the federal government formed the Office of Information and Regulatory Affairs. A primary task of the office was to bring all federal organizations into compliance with the previously defined Computer Security Act. Additionally, the office was charged with overseeing the development of functional policies, principles, standards, and guidelines. Then, in 1996, the Clinger-Cohen Act made the Office of Management and Budget (OMB) responsible for overall management of information technology (IT) investment, procurement, and security. In recognition of a growing need, the Clinger-Cohen Act also created the position of Chief Information Officer (CIO) within federal organizations.[388]

In July 1996, President Clinton further embraced protecting our critical infrastructure by creating the President's Commission on Critical Infrastructure Protection (PCCIP).[389] Beyond recognizing that threats to infrastructure existed, the commission acknowledged that the

---

[386] Cordesman, A., & Cordesman, J. (2002). *Cyber-Threats, Information Warfare, and Critical Infrastructure Protection.* Westport, CT: Praeger Publishers.

[387] Ibid.

[388] Ibid.

[389] Ibid., p. 56.

protection of the majority of U.S. critical infrastructure fell under the auspices of the private sector. According to Auerswald, Branscomb, and Michele-Kerjan, 85 percent of U.S. critical infrastructure is owned by the private sector. His work points out that market forces alone generally do not provide incentive for private sector investment in critical infrastructure protection. The PCCIP also recognized the vast scope of private sector involvement by recommending that federal officials work more closely with them.[390]

Cordesman further indicates that the PCCIP had released an initial report in October 1997. The report focused primarily on the virtual aspects of critical infrastructure vulnerabilities in lieu of physical aspects. It suggested that, as of 1997, risk levels associated with critical infrastructures were unacceptable. Technology had advanced so fast that destructive cyber tools were readily available to potential perpetrators. The report further stated that threats were numerous and diverse, and that there was a prevailing lack of awareness of the need to protect critical infrastructure. The report suggested that the protection of critical infrastructure was the responsibility of too many individuals and organizations. Hence, the report further pointed out the need for a more cohesive approach championed by a single entity or person.[391]

According to Cordesman, President Clinton followed up on the PCCIP's report by setting the stage for a new national directive entitled Policy on Critical Infrastructure Protection. In part, the directive indicated that risks were so prevalent and dangers so high that it recommended linking cyber security directly with information warfare. The policy was strongly worded and included timelines to implement and sustain U.S. critical infrastructure protection.[392] Cordesman indicates that the new policy embraced vast and far-reaching directives and actions and included establishing a new federal organization of knowledgeable

---

[390] Auerswald, P., Branscomb, L., LaPorte, T., & Michel-Kerjan, E. (2005). *The challenge of protecting critical infrastructure. Issues in Science and Technology, 22*(1), 77.

[391] Cordesman, A., & Cordesman, J. (2002). *Cyber-Threats, Information Warfare, and Critical Infrastructure Protection.* Westport, CT: Praeger Publishers.

[392] Ibid.

members representing all critical infrastructures and numerous government agencies.[393] I believe that President Clinton attempted to create universal cohesion among numerous organizations.

According to Cordesman, a member of the General Accountability Office (GAO), Mr. Jack L. Brock Jr., aired his thoughts about critical infrastructure protection before a Senate Subcommittee in February 2000. While Mr. Brock expressed appreciation for the direction of U.S. federal support, he indicated several deficit points. He expressed a vision that suggested more focus on computer security and that dated legislation and policy be recognized as such. He further questioned the validity of using the U.S. government as a role model in information security. Not necessarily questioning intent, instead, he appropriately questioned the government's ability to accomplish assigned tasks. He validated this comment using a direct link to poor security management and suggested that, in the public sector, even the simplest security measures were circumvented.[394] In defense of federal deficiencies, Janczewski proposes that management of cyber security is not trivial and includes strong information security analysis and audit. Also included are identification, mitigation, and remediation of discovered vulnerabilities.[395] In summary, Mr. Brock's vision stressed a need for far reaching information security improvement.

Computer Security and Critical U.S. Infrastructure. Protecting systems and infrastructure against cyber attack incursions is becoming increasingly difficult. Wagner suggests that today individuals possess greater power to attack computer systems. Critical infrastructure's growing reliance on computer systems and Internet interconnectivity furthers the danger.[396] Additionally, Wagner indicates that an attack on one computer system could cascade into other systems, thereby creating a domino effect.[397] Taylor suggests that several factors contribute to the United States being at particular risk to cyber attack.

---

[393] Ibid.

[394] Ibid.

[395] Janczewski, L., & Colarik, A. (2005). *Managerial Guide for Handling Cyberterrorism and Information Warfare.* Hershey, PA: Idea Group Publishing.

[396] Wagner, C. (2007). Countering cyber attacks. *The Futurist, 41*(3), 16.

[397] Ibid.

One factor is a hatred of the United States by some Middle Eastern factions. Additional potential perpetrators include anti-capitalistic governments.[398]

Auerswald brings a sense of urgency to critical infrastructure protection by saying that computer systems and connectivity are so prevalent and far reaching that attacks on critical infrastructure may possess far greater impact on public services and national security. He suggests that public and private sector lack appropriate incentives to take critical infrastructure more seriously.[399] Auerswald recommends that, in the longer term, public and private sectors must consider changes in both technologies and policies that dictate their usage and protection. He also recommends that longevity and sustainability of critical infrastructure protection is complex and blends market forces with rapid technology advancement. Creating an environment where all organizations can protect critical infrastructures requires concerted collaboration between government organizations and the private sector.[400]

Conclusion. The United States and the world have seen dramatic changes in the way people operate and maintain critical infrastructures. Many critical infrastructures share a close relationship with using computers, systems, and connectivity. If we were to eliminate these new mechanisms, supporting national security and public needs would become difficult and maybe impossible. Critical infrastructure threats loom greater and more numerous than in the past. Going forward with critical infrastructure protection requires significant investment by public and private organizations. The government must allocate resources that support more robust critical infrastructures. They must strive to create an environment that provides the private sector with investment incentives.

---

[398] Taylor, R., Caeti, T., Loper, D., Fritsch, E., & Liederbach, J. (2006). *Digital Crime and Digital Terrorism*. Upper Saddle River, NJ: Pearson Prentice Hall.

[399] Auerswald, P., Branscomb, L., LaPorte, T., & Michel-Kerjan, E. (2005). The challenge of protecting critical infrastructure. *Issues in Science and Technology, 22*(1), 77.

[400] Ibid.

Local and International Approaches to Cyberterrorism and Forensics

In this section of this chapter, I review two separate but related topics under criminal activity using computers and networks. First, I introduce *cyberterrorism* and discuss what local and international efforts are taking place to counter it. Secondly, I discuss the required steps for acquiring digital crime scene evidence. Finally, I express and stress the need to follow explicit procedures in order to qualify forensics evidence as permissible in a court of law.

What is Cyberterrorism? *Encyclopedia Britannica* categorizes the word *cyberterrorism* as a type of cybercrime. Cyberterrorism distinguishes itself from other types of cybercrime in that it attempts to disrupt, in one way or another, the very functioning of the Internet at one or more locations.[401] Janczewski references cyberterrorism as being premeditated and politically motivated.[402] Taylor struggles to define the word *cyberterrorism*. He suggests the term has been wrongly used in both academics and general media.[403] One definition Taylor cites is that terrorism is terrorism by the very nature of the act and not by the identification or underlying cause(s) of the perpetrators. He tends to view the term cyberterrorism as terror through the use of cyber tools.[404] Janczewski reports the term was first embraced first by the U.S. military in 1996 by combining the terms *cyberspace* and *terrorism*.[405]

---

[401]   Cybercrime. (2008). In *Encyclopedia Britannica*. Retrieved June 18, 2008, from *Encyclopedia Britannica Online*: *http://search.eb.com/eb/article-235699*.

[402]   Janczewski, L., & Colarik, A. (2005). *Managerial Guide for Handling Cyberterrorism and Information Warfare*. Hershey, PA: Idea Group Publishing.

[403]   Taylor, R., Caeti, T., Loper, D., Fritsch, E., & Liederbach, J. (2006). *Digital Crime and Digital Terrorism*. Upper Saddle River, NJ: Pearson Prentice Hall.

[404]   Ibid.

[405]   Janczewski, L., & Colarik, A. (2005). *Managerial Guide for Handling Cyberterrorism and Information Warfare*. Hershey, PA: Idea Group Publishing.

When we think of cyberterrorism and cyberwarfare, we think of savvy, yet malicious computer users using computers, networks, and the Internet to circumvent normal, physical barriers of entry. Once they determine how to break in, their work may become even more malignant. Their deeds range from simple website vandalism to bringing down entire critical infrastructures. Taylor suggests that web defacement activity is typically politically motivated and often becomes a tit-for-tat endeavor. Terrorists understand that some websites are highly visible and thereby provide an opportunity to spread political agendas on a global basis. Another type of malicious cyber activity is the denial of service (DOS) attack.[406] Kruse tells us that DOS attacks are not new. He reports that the Cuban government denies its citizens the ability to receive free radio broadcasts. Using the Internet to instigate DOS attacks is the latest perpetrators'venue of choice. In the cyber world, a DOS attack attempts to keep a website so busy with illegitimate calls for communication that legitimate callers cannot view the website.[407]

How are Local and International Authorities Responding to Cyberterrorism? We learn from Cordesman that as early as the 1980s, the U.S. government began taking cybercrime more seriously. The passage of the Computer Security Act of 1987 represented the first major legislation addressing cybercrime. Key issues such as federal computer policy and procedures came under intense scrutiny. With passing time, support swelled for better public and private sector collaboration in fighting cybercrime. Over the past 25 years, work has continued to address increasing requirements to protect critical infrastructures against maliciousness.[408]

The nature of the Internet and its fundamental structure create a global phenomenon. When we connect computers to the Internet, they are not bounded by national borders or geographic barriers.

---

[406]   Taylor, R., Caeti, T., Loper, D., Fritsch, E., & Liederbach, J. (2006). *Digital Crime and Digital Terrorism*. Upper Saddle River, NJ: Pearson Prentice Hall.

[407]   Kruse, W., & Heiser, J. (2002). *Computer Forensics: Incident Response Essentials*. Boston, MA: Addison-Wesley.

[408]   Cordesman, A., & Cordesman, J. (2002). *Cyber-Threats, Information Warfare, and Critical Infrastructure Protection*. Westport, CT: Praeger Publishers.

Computer connectivity transcends protections enjoyed for years by sovereign nations. As a result, an era has emerged that defines new battle grounds and attack methods employed among countries. Taylor believes information warfare should be broken into the following six categories: (1) psychological operations, (2) electronic warfare, (3) military deception, (4) physical destruction, (5) security measures, and (6) information attacks.[409]

A study by Fishman indicated a sharp increase of cyber attacks against organizations. This study of over 400 public and private companies around the world supported the notion that cybercrime is not limited to the United States alone. The estimated annual growth of attacks was a staggering 64 percent. The study showed that 70 percent of the attacks were directed at critical infrastructures such as energy, technology, and financial organizations. Over 80 percent of cyber attacks were attributed to Canada, China, Germany, Great Britain, Italy, Japan, South Korea, Taiwan, and the United States. These statistics leave little doubt that perpetrators are not hacking for mere fun. Rather, many countries are entrenched in efforts to use the Internet and global connectivity as the new battlefront.[410]

Another global community area of attention is sea transportation. International shipping of goods is both integrated with computers and dependent on connectivity. Seaports around the globe present targets of opportunity for perpetrators. A study by Shah rated U.S. seaport security from fair to poor. Such descriptions are linked to the sheer volume of global shipping activity. As such, for many nations, port security is at the top of the agenda.[411]

Russia's Vladimir Putin underscores the global attention given to cyberterrorism. The British Broadcasting Corporation (BBC) cited Putin as saying that globalization offers many opportunities but, at the same time, presents risks such as cyberterrorism. Putin placed emphasis

---

[409] Taylor, R., Caeti, T., Loper, D., Fritsch, E., & Liederbach, J. (2006). *Digital Crime and Digital Terrorism*. Upper Saddle River, NJ: Pearson Prentice Hall.

[410] Fishman, R., Josephberg, K., Linn, J., Pollack, J., & Victoriano, J. (2002). Threat of international cyberterrorism on the rise. *Intellectual Property & Technology Law Journal, 14*(10), 23.

[411] Shah, S. (2004). The evolving landscape of maritime cybersecurity. *Jamaica, 25*(3), 30.

on strengthening domestic technical organizations to promote national security.[412]

What Makes Digital Crime Scenes Different? I suggest that all crime scenes are not created equal. The information technology world is wrought with new terrain in terms of capturing, storing, and presenting evidence in a U.S. court of law. The relative newness of digital forensics and how it is handled during the discovery phase of an investigation presents tough challenges to prosecutors, defenders, and the accused. It also poses the question of whether or not current legislation keeps pace with the rate of technological advancement. In the United States, court systems develop new laws through common law, which is defined by *Black's Law Dictionary* as "the body of law deriving from law courts as opposed to those sitting in equity" (p. 294).[413] As courts resolve cases, laws arise that attempt to shape acceptable societal behavior. A lot of work still needs to be done in creating appropriate cyber-specific law.

Cordesman provides evidence that U.S. legislation does a poor job in mandating cyber security incident reporting. Some organizations do not report cyber breaches because these reports would not bode well for the organization, which may reduce both market share and stock value.[414] However, Cordesman points to the health industry as one that came under increased U.S. regulation and, hence, forced to comply. U.S. laws are clear and violations costly. A review of current legal mandates, liabilities, and penalties would provide an appropriate level of interest in cybercrime prevention, detection, and mitigation. Private sector, for-profit organizations struggle to invest in cyber security beyond what's legally required as they focus on quarterly profits.[415]

Concluding Remarks. Cyberterrorism is a new phenomenon. In the past 30 years, significant changes have transpired in computer and communications technologies. A global connectivity network provides individuals, organizations, municipalities, states, countries, and

---

[412] Anonymous (2007, July 25). Putin wants to boost domestic IT sector to counter cyberterrorism. *British Broadcasting Corporation*, pg. 1.

[413] Garner, B. (2004). *Black's Law Dictionary*. St. Paul, MN:Thomson-West.

[414] Cordesman, A., & Cordesman, J. (2002). *Cyber-Threats, Information Warfare, and Critical Infrastructure Protection*. Westport, CT: Praeger Publishers.

[415] Ibid.

national bodies with opportunities to create more efficient and effective infrastructures. This evolution is not without it danger, however. Cyberterrorism presents a clear and present danger to users of critical infrastructures. Authoritative governing bodies must catch up to these developments using financial investments and appropriate incentives to public and private sector organizations. When we understand how vulnerable and fragile infrastructures can be, we then realize that there is much more work to be done.

Protecting Critical Infrastructure: A Report to Congress

The 21st century finds the United States facing some of the same threats it has faced for many years. Standing up for democracy, free enterprise, and freedom has guided U.S. vision and action for over 225 years. During that time period, the United States was engaged in numerous confrontations. Some of those confrontations were lengthy, costly, and deadly. While these confrontations are not new to the United States, how they are being fought is new.

According to Taylor, over the past quarter century, advances in technology have created alternatives for perpetrators to force their self-serving ideologies on others.[416] As defined by Beekman, the miniaturization and convergence of telecommunication and computer equipment place great power in the hands of average citizens no matter what their agenda. Once monopolized by those with money and knowledge, computers and Internet connectivity are now common in most businesses and homes. While these changes in technology, accessibility, and affordability present positive opportunities to nations, organizations, and individuals, it brings in plain view vulnerabilities of the critical infrastructures of many countries.[417]

This chapter defines the terms *cybercrime*, *cyberterrorism*, and *cyberwarfare*. I discuss the convergence of new technologies with critical infrastructure. I then highlight protection efforts of the past and

---

[416] Taylor, R., Caeti, T., Loper, D., Fritsch, E., & Liederbach, J. (2006). *Digital Crime and Digital Terrorism*. Upper Saddle River, NJ: Pearson Prentice Hall.

[417] Beekman, G., & Quinn, M. (2006). *Computer Confluence: Tomorrow's Technology and You*. Upper Saddle River, NJ: Prentice-Hall.

how they have performed in the face of advancing technology and new vulnerabilities. I review current vulnerabilities and risk of attack on public and private critical infrastructure as well as current efforts and how they are performing. I make clear, concise recommendations on how to mitigate those risks. Finally, I propose policy recommendations for moving forward.

Introduction. To acquire a good understanding about critical infrastructure vulnerabilities and how to best mitigate them, we must first begin by placing a common framework around terms and concepts. The terms *terrorism*, *crime*, and *warfare* are familiar to most people. When we begin each of these words with "cyber," we change their meanings. Janczewski suggests that the U.S. military first coined the term *cyberterrorism* in 1996 when the term *cyberspace* was combined with *terrorism*. Additionally, when people or groups of people use computers and the Internet as tools to propagate their underlying intention(s), crime and warfare become cybercrime and cyberwarfare.[418] Taylor reports that people and organizations accomplish these activities using cyber tools. The term *cybercrime* arose from the need to describe crimes instigated through computers.[419] Taylor presents four categories of computer crime:[420]

1. Crimes where a ***computer is the target*** of a crime. Denial of service (DOS) and destroying data are two examples.
2. Crimes where ***computers are used as the instrument*** of the crime. These are crimes where a perpetrator uses a computer as an instrument to achieve some other criminal objective. Theft and exploitation are two possible examples.
3. Crimes where ***computers are incidental*** to the crime. Here the computer may not be the primary instrument of the crime

---

[418] Janczewski, L., & Colarik, A. (2005). *Managerial Guide for Handling Cyberterrorism and Information Warfare*. Hershey, PA: Idea Group Publishing.

[419] Taylor, R., Caeti, T., Loper, D., Fritsch, E., & Liederbach, J. (2006). *Digital Crime and Digital Terrorism*. Upper Saddle River, NJ: Pearson Prentice Hall.

[420] Ibid.

but rather facilitates the crime. Money laundering and child pornography are two examples.

4. Crimes by simple *association to the prevalence* of computers. Two examples here are intellectual rights and numerous types of corporate crimes.

We carry these thoughts forward as we discuss contemporary computer systems and the convergence they experience with critical infrastructure.

The Relationship between Contemporary Computer Systems and Critical Infrastructure. A reasonable level of importance and emphasis must be placed on vulnerabilities and risks associated with critical infrastructure in the United States. We learn from Beekman that contemporary computers have undergone considerable change. The point and click graphical user interface (GUI) extends the power and use of computers well beyond the technical community. Computer users no longer must possess intimate knowledge of the cryptic command-line-driven computer interface of the past. Additionally, telecommunication technology underwent substantial change over the years. The Internet adds power to computer GUIs. A complex, integrated network of communication links, the Internet spans the entire globe. Contemporary computers are designed to connect easily to the Internet. Connecting computers to the Internet around the globe provides users with near real-time information and is as commonplace today as the telephone. An analogy of the Internet might be that it is a global newspaper where anyone can sponsor information and anyone can read it.[421]

Carrying it a step further, engineers leverage the capabilities of critical infrastructure by integrating them with contemporary computer and telecommunication systems. It makes sense to combine the power of computers and telecommunications with critical infrastructure. The relationship allows infrastructure to perform more efficiently and effectively. Advanced computers and telecommunication systems are instrumental in this integration. Therefore, we have seen convergence and tight integration between computer systems and critical infrastructures.

---

[421] Beekman, G., & Quinn, M. (2006). *Computer Confluence: Tomorrow's Technology and You.* Upper Saddle River, NJ: Prentice-Hall.

This tight integration presents vulnerability and, in turn, risk. Electrical grids, water supplies, and communication networks are probably the three most affected critical infrastructures. Disrupting these types of infrastructures would present serious implications to nation states and their populations.

Nations, states, organizations, and individuals all possess the freedom to use the Internet for whatever purpose they see fit. The "openness" of the Internet elevates the opportunity for abuse by perpetrators willing to use maliciousness to support their agendas. Websites and critical infrastructures of countries are at risk of damage by such abuse. Kruse & Heiser report about commonplace DOS and web-defacing attacks. Hence, the United States must mitigate these activities by defensively posturing computers, systems, and critical infrastructure. These reasons call for the best security for U.S. critical infrastructures. To remain a sovereign nation, we must evaluate where we have been, where we now are, and where we must go in the future.[422]

Previous Programs. For some time now, the United States has addressed implications and issues surrounding contemporary technologies. Programs addressing cybercrime, cyberterrorism, and cyberwarfare began in the mid 1980s. Cordesman tells us that the U.S. government and congress took a more serious posture towards criminal cyber activity in 1987 by passing the Computer Security Act. The act was specifically focused on the need for federal government organizations to embrace newer policies and procedures on security guidelines.[423] According to Cordesman, the Computer Security Act of 1987 achieved minimal success. The expectation that all government entities would conform, including those down to the state and local levels, did not come to pass. As a consequence, the Office of Information and Regulatory Affairs was established in 1995. Additionally, in 1996, the government, through the Clinger-Cohen Act, gave the Office of Management and Budget (OMB) clear authority over computer and information security across the entire federal spectrum. At that time,

---

[422] Kruse, W., & Heiser, J. (2002). *Computer Forensics: Incident Response Essentials*. Boston, MA: Addison-Wesley.

[423] Cordesman, A., & Cordesman, J. (2002). *Cyber-Threats, Information Warfare, and Critical Infrastructure Protection*. Westport, CT: Praeger Publishers.

the position of Chief Information Officer (CIO) was created within all federal organizations.[424]

These directives guide the top leadership in federal organizations. As a result, numerous committees were created. Broad and far-reaching committee activity control many areas of computer and information security. Vulnerability and critical infrastructure protection are driving forces of these committees, which promote the need to identify specific vulnerabilities. They believe that international cooperation may provide the shortest and most sustainable route to better security. Committees have stressed the differences between public and private sector infrastructures. They highlight the fact that significantly different economic forces drive each of these sectors. Also, committees clearly expound the need to address and resolve numerous legal issues associated with implementing and using new technology. Additionally, these committees make it clear that the size and scope of critical infrastructure protection requires cooperation at all levels. All national, state, local, public, and private sectors must accept their obligations, be active, and follow through.

Current Programs. Initiatives taken during the late 20th century were excellent first steps. However, what has become clear over time is that the dynamics surrounding technology and its rate of change highlight the need to remain equally versatile with critical infrastructure protection. The 9/11 attack supports this notion the best. The 9/11 attack motivated the establishment of the Department of Homeland Security (DHS) in 2002. According to Johnson, the DHS started in early 2003 to enact the National Strategy to Secure Cyberspace. The act promotes information sharing and provides organizations with practical solutions to enhance cyber security. It also encourages organizations to protect their computer and information systems against cyber attacks. The DHS serves as the vehicle by which the United States combines numerous forums and initiatives under a single umbrella.[425]

---

[424]   Ibid.
[425]   Johnson, J.A. (2005). *Community Preparedness and Response to Terrorism: The Role of Community Organizations and Business, II.* Westport, CT: Praeger Publishers. Retrieved July 12, 2008, from *Praeger Security International Online* database at *http://psi.praeger.com.proxy1.ncu.edu/ doc.aspx?adv_search=1&term_0=cyberterrorism&index_0=words&op _1=AND&term_1=&index_1=words&op_2=AND&term_2=&index_2*

On-going congressional hearings continue to focus on protecting critical infrastructure. According to Cordesman, Mr. Jack Brock is most noted for his testimony to Congress. He makes accolades to previous work done on homeland security, but he also strongly believes that much work remains unaccomplished. Fencing infrastructure no longer protects it. Computers, information systems, and connectivity comprise the new battleground that requires protection.[426]

Policy Recommendations. There are a number of suggestions in terms of how we move forward from here with the protection of critical infrastructure. It is important to remember there are numerous facets to protecting infrastructure, and they are rich with complexities. As Cordesman indicates, the DHS has provided a substantial amount of information addressing areas which require continual attention and improvement. The following four paragraphs capture some of the most important suggestions.[427]

First, we must understand that numerous players protect our critical infrastructure. These players work in national, federal, state, local, public, and private sector organizations. With such wide geographic dispersion and decentralization, all players must clearly understand the United States' plan to protect critical infrastructure. The DHS provides this across-the-board understanding and centralized focus. Each player must understand terminologies, concepts, and philosophical approaches used by the United States. Additionally, each organization must understand its roles to protect critical infrastructure. All players must assume their responsibilities and obligation. In building a common understanding framework, the players must identify vulnerabilities and calculate risks inherent in the most critical infrastructures. All participating organizations must clearly understand that critical infrastructure protection is a task too vast for any single organization to do alone. It requires all organizations to work synergistically together.

=words&op_3=AND&term_3=&index_3=words&s=r&freeform=&d=/ books/dps/2000c277/2000c277-p2000c2779970041001.xml&i=0.

[426] Cordesman, A., & Cordesman, J. (2002). *Cyber-Threats, Information Warfare, and Critical Infrastructure Protection.* Westport, CT: Praeger Publishers.

[427] Ibid.

Second, the U.S. government must place the most talented individuals in the most crucial roles within the federal protection programs. The government must hire, train, and retain talented professionals in order to best create the next-generation infrastructure systems, which minimize vulnerability and possess manageable risks. Hence, the government must be creative and offer attractive incentives to qualified professionals. Additionally, these professionals must receive genuine support from their leaders in order to accomplish their formidable challenges.

Third, professionals working on critical infrastructure protection must apply stronger program/project management mindsets. These huge efforts must be worked using the best minds and best available tools to succeed. A talented pool exists of professionals who possess extensive program/project management experience. Organizations working on critical infrastructure protection must seek and recruit these people.

Fourth, an array of legal issues applies to computers used in the cyber arena. Because the Internet does not recognize sovereign borders, it creates new issues. The connectivity that the Internet provides to critical infrastructure and the ability to affect these infrastructures from anywhere in the world presents new challenges to legal professionals. We must evaluate all aspects of legal issues on a continuing basis. Laws governing computer and connectivity issues are currently obsolete. We must resolve this inadequacy soon.

Concluding Remarks. Undoubtedly, the United States and the world must recognize this new battleground, which is far different than previous war zones. Critical infrastructures, information systems, computers, telecommunications, and all operations and controls have converged. It's a modern marvel how they integrate and perform in such a transparent and seamless fashion. Mankind relies on critical systems to perform flawlessly. People who inflict suffering on the masses by damaging/destroying critical infrastructure exist in the world. The United States and other nations must understand critical infrastructure vulnerabilities. However, a clear understanding is just the start. As we move forward, focus must be given to protecting and sustaining critical infrastructures and the reliable services they provide.

Chapter 13
Security of Handheld Devices

Rhonda G. Chicone, PhD (ABD)

T his chapter describes the necessity of securing mobile wireless devices. It outlines my concern that mobile wireless devices represent a new target for cybercriminals. Many large and small, private and public organizations today do not implement adequate mobile device security to ensure safe and secure information. However, all is not lost because methods and technologies exist that could help. In this chapter, I also outline some guidelines to implement an effective mobile wireless device security strategy for any size organization. Furthermore, I define mobile wireless devices and mobile application software systems that power them strictly as handheld devices, which include mobile phones, personal digital assistants (PDAs), and smart phones (but excludes laptops, for example).

A New Target for Cybercriminals

Mobile devices and communications networks allow information to be accessed at any time and from anywhere. This section focuses on my concern that the managers of many organizations today do not understand mobile wireless device hardware and software technology well enough to implement appropriate security measures. Without the appropriate understanding and knowledge, successful security measures cannot be implemented. Without the proper security measures in place, there are potentially enormous information security gaps. Security gaps, that if not filled, can become a tremendous opportunity for cybercriminal activities. The potential security threats could affect organizations of any size, public or private, that permit mobile devices, and the associated software that drives them, within the organization's walls.

Introduction. Every day, we see new television commercials, Internet ads, or billboards introducing the world to the next *killer* mobile wireless device. Mobile wireless devices come in many forms or classifications, e.g., PDAs, cell phones, smartphones, and camera phones. These feature-rich devices work on high-speed cellular networks and 802.11 wireless fidelity (WiFi) networks. In 2008, mobile phone vendors shipped 1.18 billion units worldwide.[428] Consequently, the number of business professionals using mobile wireless devices in the workplace grows exponentially. Recent research indicates that there will be over 269 million enterprise users of mobile devices by 2010.[429] Organizations face many security threats. Therefore, information technology (IT) professionals must understand fully mobile technology and available security options. Additionally, mobile device users should participate in training and awareness programs. Unfortunately, most employees and managers do not realize what they will lose by not incorporating security controls in their mobile devices. Furthermore, security managers or chief information officers (CIOs) at many organizations do not know what information they possess, where it is located, or even its worth.[430]

Why Use Mobile Devices in the Work Place? According to officials at the International Data Corporation,[431] more than 70 percent of the U.S. workforce will own mobile devices by the end of 2009. Korn/Ferry International reported that 81 percent of executives,

---

[428] International Data Corporation. (2009). *Worldwide mobile phone market declines by 12.6% in fourth quarter, more challenges to come says IDC*. Retrieved February 8, 2009, from *http://www.idc.com/getdoc. jsp?containerId=prUS21659209*.

[429] MacLachlan, G. (2007). *Protecting the wireless enterprise*. Retrieved October 5, 2007, from *http://info4security.com/story.asp?sectioncode=1 0&storycode=4114413&c=1*.

[430] Beaver, K. (2007). *Mobile security: Top oversights*. Retrieved October 10, 2007, from *http://searchmobilecomputing.techtarget.com/tip/0,289483, sid40_gci1271132,00.html*.

[431] International Data Corporation. (2007). *A guide to delivering dynamic protection in an evolving threat environment*. Retrieved November 20, 2007, from *http://viewer.bitpipe.com/viewer/viewDocument. do?accessId=6837682*.

globally, are now constantly connected via mobile devices.[432] Clearly, the way people work has changed and will continually change as mobile technology advances at such a rapid rate. Whether IT departments or analogous groups within organizations know what mobile technology to secure, how to secure it, and how to maintain it, remains an open question.

In 2008, several factors drove the use of mobile devices. These factors include lifestyle choices, productivity gains, and technology improvements.[433] Businesses continue to witness significant productivity benefits by keeping workers functional with mobile devices. Researchers and analysts at the Economist Intelligence Unit Limited polled more than 500 global businesses in more than 200 countries in eight key industries and asked how their organizations use business mobility to be more agile. They found that competitive factors push businesses toward greater mobility as business mobility enters the mainstream.[434]

While away from their office, users can manage their work calendar from a mobile device or review a presentation or military plan of action and provide comments via e-mail—all from the palm of their hand. Hence, mobile devices, like personal computers, represent an asset to any enterprise. Consequently, the trend towards total access creates a need for increased security and privacy of organizational data available on or via mobile applications and devices—security that is unavailable in many organizations.[435]

---

[432] Hickey, A. R. (2007). *Mobile security breaches inevitable, study says.* Retrieved October 5, 2007, from *http://searchmobilecomputing.techtarget. com/originalContent/0,289142,sid40_gci1272948,00.html.*

[433] Friedman, J., & Hoffman, D. V. (2008). Protecting data on mobile devices: A taxonomy of security threats to mobile computing and review of applicable defenses. *Information Knowledge Systems Management, 7,* 159-180.

[434] Economist Intelligence Unit Limited. (2007). *The quest for competitiveness: Business mobility and the agile organization.* Retrieved November 27, 2007, from *http://viewer.bitpipe.com/viewer/viewDocument. do?accessId=6852945.*

[435] Tarasewich, P., Gong, J., Nah, F. F., & DeWester, D. (2008). Mobile interaction design: Integrating individual and organizational perspectives. *Information Knowledge Systems Management, 7,* 121-144.

Capabilities of Mobile Devices. Do not be misled—mobile devices are, in fact, *computers* that fit in the palm of your hand. Some mobile devices possess more processing power and storage then PCs and servers of just a few years ago. To put their power in perspective, a company I know had replaced one of its production Web servers about a year ago. The processing power of the servers was 800 megahertz, and random access memory (RAM) was 256 megabytes. Similarly, a new smartphone made by Apple Inc., that came to market in June 2009, called the iPhone 3GS, has a 600-megahertz microprocessor and 256 megabytes of RAM.[436] Hence, today's mobile devices can store and process enormous amounts of confidential information.

Mobile devices include strong processing power, improved storage, superior battery life, fast wireless data transfer rates, and robust functionality created by integrating hardware and software. Mobile devices possess almost everything a business professional would need in one device. They include a phone, voicemail, notepad, e-mail, calendar, text messaging, Internet browsing, and contact management. Mobile computing continues to provide organizations in numerous industries with an array of information system implementations because hundreds of mobile applications were developed for business use.[437] However, adopting, deploying, securing, and maintaining this type of mobile technology presents many challenges. Mobile device security challenges require focus and attention and cannot be an afterthought because mobile devices represent an easy target for cybercrime.

Examples of Mobile Application Software. Electronic mail, known as e-mail, is a software application fundamental to using the Internet. Corporate e-mail is delivered wirelessly to a mobile device with special e-mail delivery software installed in the mobile device. Software is also present on the computers behind corporate firewalls.

---

[436] Gadget Review, Inc. (2009). Retrieved July 1, 2009, from *http://www. gadgetreview.com/2009/06/iphone-3gs-processor-speed-and-ram-published-by-t-mobile-netherlands.html.*

[437] Hu, W. C., Zuo, Y., Wiggen, T., & Krishna, V. (2008). Handheld data protection using handheld usage pattern identification. *Proceedings of the 2008 IEEE International Conference on Electro/Information Technology, Ames, Iowa, 1,* 227-230.

The software behind the corporate firewall and the software on the mobile device work together to achieve wireless e-mail delivery.[438] Proprietary information, such as customers' personal information or governmental political plans, is transmitted to and from mobile devices using this special e-mail delivery software.

Location-based services represent another example of a mobile application. This application type provides nearby points of interest based on the real-time location of the mobile user. However, locating or tracking a mobile user raises several privacy and security concerns.[439] For example, a wireless subscriber of a location-based service may be concerned about their personal safety if their location were to fall into the wrong hands. Moreover, location could infer other personal information such as political or religious affiliation, state of health, or personal preferences.

Security Breaches. According to the Ponemon Institute, an organization that conducts independent research on privacy, data protection, and information security policy, the average cost of a data breach in 2008 had reached $202 per compromised customer record compared to $197 per record in 2007. The source of many of these data breaches were portable devices like laptops and flash drives.[440] Considering all of this, a high probability exists that a CEO or congressman one day will lose his/her mobile device or have it stolen, and the thief uses the information on the device for criminal purposes.

Problems that arise because of inadequate security measures for mobile application software include unauthorized use of mobile devices, reckless behavior, theft, viruses, and spyware. Depending on the organization and compromised data type, damage caused by any of these security breaches would be costly. International Data Corporation analysts assert

---

[438] Research in Motion Limited. (2006). *BlackBerry Enterprise Solutions security.* Retrieved September 5, 2008, from *http://na.blackberry.com/ eng/ataglance/get_the_facts/xHCO-BlackBerry_Enterprise_Solution_ Security_version_4.pdf.*

[439] Atluri, V., Shin, H., & Vaidya, J. (2008). Efficient security policy enforcement for the mobile environment. *Journal of Computer Security, 16*, 439-475.

[440] PGP Corporation. (2009). *2008 annual study: Cost of a data breach.* Retrieved April 1, 2009, from *http://www.pgp.com/insight/research_reports/.*

that information protection and control will continue as a substantial security investment over the next five years. More insider scandals such as leaking of customer records, confidential information, and intellectual property will continue to be a problem.[441] However, designing technology solutions helps prevent deliberate or inadvertent disclosure of sensitive information for organizations. Specifically, organizations should create or improve solutions that monitor, secure, encrypt, filter, and block sensitive information in data at rest and in motion.

Although not a security breach involving a mobile device, what happened at TJX on January 17, 2007 (when an unauthorized intrusion into its computer systems had occurred), illustrates a good example of the vulnerability of many organizations. The computer system stores information related to customer transactions that includes credit cards, debit cards, check cards, driver's licenses, and merchandising for customers of T. J. Maxx, Marshalls, HomeGoods, and other retail stores in the United States, Puerto Rico, and Canada.[442]

Alfred Loo, associate professor of computer science at Lingnan University, found that some corporate security teams believe that mobile phones were for personal use only, and it was not their duty to protect the applications on them. However, Loo contends that mobile phones represent the next easy target for professional hackers. He found that security teams generally assume that smartphones possess the necessary security features already within the device. Security teams also assume that these and similar phones maintain enough security features that hackers show no interest in them, that no viruses

---

[441] International Data Corporation. (2007). *Targeting inside threats. International Data Corporation white paper.* Retrieved November 20, 2007, from *http://viewer.bitpipe.com/viewer/viewDocument.do?accessId= 6840322.*

[442] TJX. (2007). *The TJX Companies, Inc. victimized by computer systems intrusion: Provides information to help protect customers.* Retrieved November 15, 2007, from *http://www.businesswire.com/ portal/site/tjx/index.jsp?epi-content=GENERIC&newsId=20070117 005971&ndmHsc=v2*A938775600000*B1196043128000*C410249 1599000*DgroupByDate*J2*N1001148&newsLang=en&beanID= 1809476786&viewID=news_view.*

infect smartphones, and that little would be lost should hackers attack smartphones anyway.[443]

Security Prevention Starts at the Top. In response to a 2007 survey, CIOs indicated that they thought that mobile security breaches were inevitable over the next five years. Furthermore, 86 percent of CIOs surveyed thought that solving data security problems that mobile devices pose to their organizations is one of the most significant issues that they face.[444] Profit-motivated attackers target business users as an avenue into corporate data centers. The largest target of opportunity motivates cybercriminals, and the rapid increase in mobile device use in the workplace inevitably draws attention to these vulnerable mobile systems.[445]

Unfortunately, there is no one-size-fits-all solution. However, the first step to battle the mobile device security problem and the potential for criminal activity is to establish a security policy endorsed by the organization's top management. The organization's Security department must then enforce, monitor, and periodically review the security policy. However, people represent a critical element in effectively securing an organization. Therefore, top management should implement and endorse effective training and awareness programs. Unfortunately, even with the most comprehensive awareness programs, unpredictable people commit human errors that comprise 42 percent of all data breaches.[446] Finally, the group responsible for an organization's mobile device strategy must

---

[443] Loo, A. (2009). Security threats of smart phones and bluetooth. *Communications of the ACM, 52*(3), 150-152.

[444] Hickey, A. R. (2007). *Mobile security breaches inevitable, study says.* Retrieved October 5, 2007, from *http://searchmobilecomputing.techtarget. com/originalContent/0,289142,sid40_gci1272948,00.html.*

[445] Friedman, J., & Hoffman, D. V. (2008). Protecting data on mobile devices: A taxonomy of security threats to mobile computing and review of applicable defenses. *Information Knowledge Systems Management, 7,* 159-180.

[446] International Information Systems Security Certification Consortium. (2007). *Securing the organization: Creating a partnership between HR and information security.* Retrieved December 9, 2007, from *https://www. isc2.org/Documents/HiringGuide/HRWhitePaper.pdf.*

understand what security options exist so that they can implement, verify, and monitor the appropriate options.

Good News on the Horizon? In a study of more than 1,000 IT professionals, including data from 3,600 surveys completed since 2002, 78 percent indicated that management now considers information security a top priority. As a result, the majority of these companies have developed a comprehensive written information security policy. Among those with a written policy, 81 percent indicated that their policy included specific information to cover *mobile employees*. In addition, some organizations made security-related training for their information security staff mandatory—especially for new hires. Among the organizations that have security training, 80 percent believe that the training had improved security at their organization.[447]

Concluding Remarks. Mobile devices and communications networks allow access to information at any time and from anywhere. Organizations are rapidly adopting mobile device technology and seeing major productivity benefits of keeping workers fully functional when mobile. However, cybercriminals target the business user as a passageway into corporate data centers.[448] Thus, security for mobile devices and the software that drives them become critical in protecting organizational information.[449]

A Proposed Guide for Implementing Handheld Security

Protecting an organization's information remains crucial in today's digital information age. An organization's proprietary information may reside in a mainframe computer, personal computers, laptops,

---

[447] Computing Technology Industry Association. (2007). *Trends in information security: Analysis of IT security and the workforce*. Retrieved December 9, 2007, from   *http://www.comptia.org/sections/research/white%20papers/ SecurityWhitePaper0507.pdf.*

[448] Hickey, A. R. (2007). *Mobile security breaches inevitable, study says*. Retrieved October 5, 2007, from *http://searchmobilecomputing.techtarget. com/originalContent/0,289142,sid40_gci1272948,00.html.*

[449] Tauschek, M. (2008). *Mobility policies: Security*. Retrieved May 5, 2008, from   *http://viewer.bitpipe.com/viewer/viewDocument.do?accessId= 7500007.*

and mobile wireless handheld devices. Handheld devices are complex computers in their design and architecture as are the wireless networks in which they work.[450] Using handheld devices without the correct security preventive measures in place exposes an organization to risks. These risks could cause a security breach or attack, which could cost an organization a tremendous amount of money. However, securing an environment brings accompanying technical and financial challenges. An unrealistic goal would be to make your organization fully secure. Because technology changes so rapidly, it makes 100 percent security nearly impossible to accomplish. Nevertheless, an organization can take steps to mitigate security risks and reduce the likelihood of a security breach with a mobile device or a system breach caused by a mobile device.

Key Terms You Should Know. Before moving to the proposed guidelines for handheld security, I define here some key terms applicable to the topic of handheld security:

- *Authentication*. A mechanism for identifying a user to a computer network. For mobile devices, it is a mechanism for identifying the device to a computer network.[451]
- *Bluetooth*. An open wireless protocol for exchanging data over short distances from fixed and mobile devices, creating personal area networks (PANs).[452] Most handheld devices on the market today have built-in Bluetooth capabilities.
- *Computer Worm*. A computer program that self replicates. A computer worm can use a computer network to send copies of itself to other computers on the network, including mobile devices.[453]

---

[450] Shih, D., Lin, B., Chiang, H., & Shih, M. (2008). Security aspects of mobile phone virus: A critical survey. *Industrial Management & Data Systems, 108*(4), 478-474.

[451] Melnick, D., Dinman, M., & Murator, A. (2003). *PDA security: Incorporating handhelds into the enterprise*. New York: McGraw-Hill.

[452] Ibid.

[453] Shih, D., Lin, B., Chiang, H., & Shih, M. (2008). Security aspects of mobile phone virus: A critical survey. *Industrial Management & Data Systems, 108*(4), 478-474.

- *Data Encryption.* A method for encoding data into unreadable format while transmitting it from a sender to a recipient.[454] It is a secondary level of protection after authentication.
- *Data-in-motion.* Sometimes called data-in-transit. These data are traveling from point A to point B. For example, data transferring through the Internet.
- *Data-at-rest.* These are the data located on a mobile device.
- *Denial-of-service.* Attacks that keep trying to exhaust a device's battery and keep the device active, rendering the device useless.[455]
- *Encryption Algorithm.* Complex mathematics used in conjunction with an encryption key to create encrypted data.[456]
- *Encryption Key.* A number, word, or phrase used in an encryption algorithm.[457]
- *Encryption Key Length.* Encryption key lengths vary in size. Longer keys will help protect against a hacker decoding the key.[458]
- *Firewall.* A hardware device or software that protects an environment by analyzing network data and determining what is permitted to pass through based on a set of rules.[459]

---

[454] Osayamwen, E. (2004). Encryption key management strategy. (Doctoral dissertation, Northcentral University, 2004). *Dissertation Abstracts International, 65,* 3005.

[455] Shih, D., Lin, B., Chiang, H., & Shih, M. (2008). Security aspects of mobile phone virus: A critical survey. *Industrial Management & Data Systems, 108*(4), 478-474.

[456] National Institute of Standards and Technology. (2001). *Federal Information Processing Standards 140-2. Security requirements for cryptographic modules.* Retrieved March 20, 2008, from *http://csrc.nist. gov/publications/PubsFIPS.html.*

[457] Osayamwen, E. (2004). Encryption key management strategy. (Doctoral dissertation, Northcentral University, 2004). *Dissertation Abstracts International, 65,* 3005.

[458] National Institute of Standards and Technology. (2001). *Federal Information Processing Standards 140-2. Security requirements for cryptographic modules.* Retrieved March 20, 2008, from *http://csrc.nist. gov/publications/PubsFIPS.html.*

[459] Melnick, D., Dinman, M., & Murator, A. (2003). *PDA security: Incorporating handhelds into the enterprise.* New York: McGraw-Hill.

- *Malware.* A short term for malicious software. Software that is designed to infiltrate or damage a computer system without the owner's informed consent.[460]
- *Middleware.* Software application(s) used in an organization that connects to other software applications within the organization. The applications may then exchange data to achieve the intended goal of the middleware application.[461]
- *Mobile Virus.* A software application that targets mobile phones or wireless-enabled PDAs. The first mobile phone virus, called VBS. Timofonica, was identified on May 30, 2000.[462]
- *Mobile Security Policy.* A document or the part of an information security policy that describes how mobile device information will be secured and kept private within the organization.[463]
- *Port.* A physical or logical pathway along which computer data can flow.[464]
- *Secure Socket Layer (SSL).* A secure communications channel that uses public key cryptography for a client to authenticate to a server.[465] SSL is mainly seen when a browser on a PC or mobile device accesses a Website and the Web address is encoded with HTTPS.

---

[460] Shih, D., Lin, B., Chiang, H., & Shih, M. (2008). Security aspects of mobile phone virus: A critical survey. *Industrial Management & Data Systems, 108*(4), 478-474.

[461] Brans, P. (2003). *Mobilize your enterprise: Achieving competitive advantage through wireless technology.* Upper Saddle River, NJ: Prentice Hall.

[462] Shih, D., Lin, B., Chiang, H., & Shih, M. (2008). Security aspects of mobile phone virus: A critical survey. *Industrial Management & Data Systems, 108*(4), 478-474.

[463] Brans, P. (2003). *Mobilize your enterprise: Achieving competitive advantage through wireless technology.* Upper Saddle River, NJ: Prentice Hall.

[464] Melnick, D., Dinman, M., & Murator, A. (2003). *PDA security: Incorporating handhelds into the enterprise.* New York: McGraw-Hill.

[465] Brans, P. (2003). *Mobilize your enterprise: Achieving competitive advantage through wireless technology.* Upper Saddle River, NJ: Prentice Hall.

Guidelines for Best Practices. I recommend four fundamental guidelines for securing a mobile environment. First, the highest level of the organization should define and endorse a mobile device security policy. Second, the IT group or department responsible for handheld deployment and support should receive adequate mobile security training. Third, as with the IT/security group, those using the handheld devices should receive appropriate training and awareness. Finally, to ensure successful implementation of the mobile device security policy, the IT/security group must understand the possible types of mobile security breaches and threats and the available mobile security options.

Mobile Device Security Policy. An organization of any size can implement a mobile device security policy, which typically is a subset of the organization's larger information security policy. The policy outlines the requirements and rules that must be followed when using mobile devices within the organization. The policy should be written to contain sufficient definitions of what must be done to protect mobile information and a listing of people using handheld devices in the organization. The objective of a well-written and successfully implemented security policy is to improve information availability, integrity, and confidentiality from both inside and outside the organization.[466]

To ensure success and the acceptability of the mobile device security policy in the organization, someone within the organization must "own it." This person should be of such a status in the organization to ensure proper implementation of the mobile security policy and continual review for improvement. Therefore, the organization should appoint an information security officer (ISO). The ISO could be an existing person in the organization or someone new to the organization. The ISO must lead the way to assure that the organization delegate proper authority to stakeholders, technical personnel, and decision makers to enforce the policy.[467] The SANS (SysAdmin, Audit, Network, Security) Institute,

---

[466] SANS Institute. (2001). *Security policy roadmap - Process for creating security policies*. Retrieved July 9, 2009, from *http://www.sans.org/reading_room/whitepapers/policyissues/security_policy_roadmap_process_for_creating_security_policies_494?show=494.php&cat=policyissues.*

[467] SANS Institute. (2001). *Security policy roadmap—Process for creating security policies*. Retrieved July 9, 2009, from *http://www.sans.org/reading_room/whitepapers/policyissues/*

located at *http://www.sans.org*, is a valuable resource for understanding security policies and how they are created.

IT Training. Handheld devices with new technological bells and whistles come to market every day. What are the security implications of these bells and whistles? Could they cause a security breach or new pathway for an attack with the mobile device as the weapon? Those responsible for ensuring a secure mobile environment should receive on-going training. This training should be mandatory by every organization that allows handheld devices for worker activities. Many believe that they have nothing to lose, but they should think again. A security breach or attack could be very costly, much more so than the cost to implement necessary training within the organization. One way to implement necessary training is to allow the IT/security group time each month to surf applicable handheld security websites to learn about new developments. Additionally, universities offer many inexpensive online training courses. Finally, for members of the Association of Computing Machinery (www.acm.org), many online classes are available free upon payment of membership fees.

End User Training. The organization must ensure that all end users fully understand the mobile security policy. First, the organization should distribute copies of the policy and require all users to read it. Additionally, all mobile users should receive policy training that details roles and responsibilities, rights and obligations, and repercussions for violations. Lastly, after policy training, users should sign off that they understand and agree to abide by the policy[468]. I highly recommend conducting periodic "how-to" training classes. Finally, awareness newsletters and/or awareness posters represent inexpensive ways to educate users on the security and use of handheld devices.

Existing Mobile Security Solutions or Methods. The IT department or group responsible for mobile device security could apply many handheld device strategies. These strategies include using strong

---

security_policy_roadmap_process_for_creating_security_ policies_494?show=494.php&cat=policyissues.

[468] Tauschek, M., & Angl, J. (2008). *Mobility policies: Responsibility and education*. Retrieved May 5, 2008, from *http://viewer.bitpipe.com/viewer/ viewDocument.do?accessId=7500007.*

passwords, changing passwords frequently, encrypting data in motion, enabling data-at-rest encryption either native to the device or with third-party software, installing and updating virus protection, or using a mobile firewall.[469] In addition, mobile device management software, a type of middleware, is available on the market to help IT/security groups protect their mobile users. Depending on the vendor, mobile device management software possesses a range of capabilities. For example, mobile device management software helps IT/security personnel to remotely remove all data (also known as a *kill pill)* from a device if that device is lost or stolen. Also, mobile device management software helps the user to define strong passwords and to force password changes automatically, thereby protecting the negligent user. Notify Technology Corporation, located at *www.notifycorp.com*, is a good resource for mobile device management software.

Concluding Remarks. I hosted a meeting in June 2009 where the topic of the meeting was a discussion about mobile device security. The people in the meeting were CEOs and CTOs of organizations both outside and within the United States. All those present were very knowledgeable about mobile device applications and mobile device security. Their companies use mobile devices, and companies they support or sell to also use mobile devices. The organizations they do business with range from very small to very large enterprises including government organizations. Everyone agreed that most organizations, unless they must be compliant due to regulatory requirements, implement mobile security measures as a *reactionary* rather than a *proactive* function. Additionally, many agreed that something major and newsworthy must *first* happen before organizations will truly understand the consequences and take things seriously. For example, one person in the meeting felt that mobile device security will not be taken seriously until someone hacks President Obama's BlackBerry and leaks confidential information.

Protecting an organization's information remains crucial in today's digital information age. In their design and architecture, handheld devices represent complex computers as do the wireless networks in which they

---

[469]   Halpert, B. (2004). Mobile device security: In *Proceedings of the First Annual Conference on Information Security Curriculum Development* (pp. 99-101). New York: Association for Computing Machinery.

operate.[470] Using handheld devices without correct security prevention measures exposes an organization to risks that could potentially cause an expensive security breach or cyber attack. By implementing these guidelines, organizations can take the necessary steps to protect handheld mobile devices, users, and proprietary information.

---

[470] Shih, D., Lin, B., Chiang, H., & Shih, M. (2008). Security aspects of mobile phone virus: A critical survey. *Industrial Management & Data Systems, 108*(4), 478-474.

Chapter 14
Summary

Robert T. Uda, MBA, MS

W e might just as well get used to it. Now with us include cybercrime, cyberterrorism, and cyberwarfare, and they will be around for a long time. With the advent of the computer, networks, and the Internet, regular crime has transitioned into cybercrime, ordinary terrorism now includes cyberterrorism, and conventional warfare now incorporates cyberwarefare. In this chapter, I have summarized all of the previous chapters in this book for easy reference.

Cybercrime, Cyberterrorism, and Cyberwarfare in Perspective

Cybercrime is with us on a daily basis (24 hours a day and 7 days a week). Furthermore, it will continue on as long as we use computers, networks, phones, videos, webcams, emails, and the Internet. However, the most deadly one of the three is cyberwarfare. Cyberwarfare is not being conducted against us by individual, civilian hackers or by terrorist organizations but is being conducted by nation states in an organized, methodical manner.

The United States must be prepared to conduct cyberwarfare. Indeed, we must be ready to conduct offensive cyberwarfare instead of just defensive cyberwarfare. Additionally, we must be prepared for massive cyber retaliation should our entire nation be massively attacked by China, Russia, Korea, Iran, or any other nation that would do us harm. Massive Retaliation was a national strategy that ended with the close of the Cold War and the demise of the Soviet Union. However, it should be revived for cyberwarfare, or we may see the demise of the United States. We must not allow that outcome to happen.

Domestic Efforts to Combat Cyberterrorism

It is only a matter of time before cyberterrorists perpetrate an attack on a city or county government. All manmade and natural disasters are local events with local units being the first to respond and the last to leave. However, the state government must be the driving force behind the efforts of defining what needs to be done, establishing a coordinated statewide strategy, and performing a comprehensive threat assessment. Sharing information to protect systems is an important foundation for ensuring governmental continuity. Partnering is a key mechanism in creating a closer working relationship among, national, state, and local governments. If these things are done, we would have a much more effective effort in combating cyberterrorism at the state and local governmental levels.

Cyberterrorist threats are on the rise. The U.S. government is not keeping close surveillance on the cyberterrorism threat. Consequently, in the current threat environment, the private sector remains the first line of defense for its own facilities. Furthermore, since the 9/11 attacks, nationwide enterprises have increased their investments in security to meet the demands of the new threat environment. Hence, rather than governments and the military carrying the protection load, the burden of watching and preparing for a cyber attack on U.S. CIs rests on the private sector. Therefore, private industry has taken a proactive approach to combat cyberterrorism, which is a good thing.

International Efforts to Combat Cyberterrorism

The United States is making combined international efforts to combat cyberterrorism. This approach is an important and sure way to beat cybercriminals and cyberterrorists. Other countries of the world use cyber protection measures that the United States could adopt. If the United States will do these things that we are not currently doing or are not doing well, we should do them to be more effective in fighting the Global War on Terrorism (GWOT). I believe that some of these measures could be effective applied here in the United States. If the United States will add these protective measures to our present repertoire of tools, we will be even more effective in combating cyberterrorism.

U.S. Policy to Prevent a Cyber Attack

The Electronic Communications Privacy Act does not curtail cyber attackers. Generally, companies, government agencies, and academia are inadequately prepared. We pay too little attention to security. We devote too few resources to it. Management needs to make security a priority. CEOs and boards of directors need to pay attention to security and make sure resources are devoted to it.

Under the principle of *nullum crimen sine lege*, the legal system cannot prosecute a perpetrator unless the law prohibits his/her behavior (no matter how harmful it may be). Hence, the Philippine government did not prosecute the perpetrator of the "I LOVE YOU" virus because no law existed that prohibited the release of malicious code.

It is important that companies preserve their computer-stored data for the required duration of time. Due to the ease with which perpetrators can manipulate computer data, the court could likely reject the data as evidence if a trained computer forensics specialist does not perform the search/analysis. It is important that Internet service providers (ISPs) help law enforcement to conduct joint investigations. The stakes in the "cat and mouse" game is getting higher. Spam and cybercrime are really about the money. There are people making a lot of money out there. To curtail cybercrimes, there must be a unified effort between government and the private sector.

Effectiveness of U.S. Infrastructure Programs

Technology makes life easier and more efficient for all of us. However, computer crime threatens our commercial and personal safety. Computer forensics has become an indispensable tool for law enforcement. In the digital world, as in the physical world, the goals of law enforcement must balance with the goals of maintaining personal liberty and privacy. American criminal investigators have wrestled with these same issues for over 200 years. Currently, we lack a national framework for curricula and training development with no gold standard for professional certification.

Cybercrime is a major international challenge. We need a collective security approach to protect the global critical infrastructure (CI). The Convention on Cybercrime, which became effective in July 2004, is the *first and only* international treaty to deal with breaches of law over the Internet or other information networks. Policymakers and the public today see the threat of a terrorist attack on information technology (IT), or *cyberterrorism*, as being one of the greatest dangers to the United States. Cyber security is a community issue. By keeping our eyes open and applying effective techniques, organizations will prevent attacks and recover quickly after an attack. However, we still are not where we must be to possess effective infrastructure programs in the United States.

Critical Infrastructure Protection Program

Terrorists seek to undermine confidence in our public and private institutions and in our ability to manage the consequences of their attacks. In response, the federal government must work collaboratively and in partnership with state and local governments, with the private sector, and with local citizens. To the extent that government and private industry are believed to be doing everything within reason to protect the public from harm, the public's confidence in its institutions will remain intact despite such attacks. We can have the best national strategy for homeland security that the most brilliant minds in Washington can devise, and yet we will fail in our endeavor if local communities do not meet the immediate challenges of a terrorist disaster.[471]

Local governments represent the front lines of protection and the face of public services to the American people. Their core competencies must include knowledge of their communities, residents, landscapes, and existing critical services for maintaining public health, safety, and order. Communities look to local leadership to assure safety, economic opportunities, and quality of life (QOL). Public confidence, therefore, starts locally and is dependent upon how well communities plan and

---

[471] Juster, K. I. (2002, February 13). Homeland security and critical infrastructure assurance: The importance of community action. Remarks of the Undersecretary of Commerce for Export Administration at the conference on Critical Infrastructures: Working Together in a New World, Austin, Texas. Retrieved April 18, 2008, from *http://www.bis.doc.gov/ news/2002/communityactionimportantnhomelandsecurity.htm.*

protect their citizens, respond to emergencies, and establish order out of chaos.[472] To this end, the City of North Miami Beach Police Department, the private sector, and concerned citizens have begun an important partnership and commitment to action.[473]

Recommendations on Cyber Threats and Warfare

One thing is for sure: success in future conflicts will depend less on bombs and bullets and more on bits and bytes.[474] In the end, the cyber threat is revolutionary, officials said, because it has no battle lines, the intelligence is intangible, and attacks come without warning leaving no time to prepare defenses. Education and training of computer users, not enforcement, are the most effective defense measures, officials said.[475]

If our enemies conduct a cyber war against us, we must be prepared to defend ourselves and also be capable of taking the offensive. *The best defense is a good offense.* Little is known about the precise nature of Washington's offensive capabilities. State—or group-sponsored attacks against our information systems using computer viruses and other techniques should be considered an act of war. Tightly coupling business and industry into the cyber war defense strategy is arguably the most critical component. It represents the one area that the government has the worst track record, which must be improved. As a war-fighting domain, cyberspace favors the

---

[472] Anonymous (2003, February). *The National Strategy for the Physical Protection of Critical Infrastructures and Key Assets.* Washington, DC: The White House.

[473] Anonymous(n.d.).CityofNorthMiamiBeach,Florida,CriticalInfrastructure Protection Program. The official website of the City of North Miami Beach, Florida. Retrieved April 18, 2008, from *http://northmiamibeach. govoffice.com/index.asp?Type=B_BASIC&SEC=%7BF04CACB1-1126-4FF5-85C3-81ACF84754F9%7D.*

[474] Coleman, K. (2007, November). Department of Cyber Defense: An organization whose time has come! The Technolytics Institute, 7 pp.

[475] Rogin, J. (2007, February 13). Cyber officials: Chinese hackers attack 'anything and everything.' *FCW.com.* Retrieved May 3, 2008, from *http:// www.fcw.com/online/news/97658-1.html?type=pf.*

offense. The first battle of the next war will be fought and won in the cyberspace arena.

Strategy to Combat Cyberterrorism

There is no uniform consensus on a universal definition of the word *cyberterrorism*. We would be hard pressed to develop a definition that every nation in the UN would agree upon. So, we move forward anyway.

Frank Cilluffo, an analyst at the Center for Strategic and International Studies in Washington, DC, testified to the Senate Government Affairs Committee in October 2001. He said, "Bits, bytes, bugs, and gas will never replace bullets and bombs as the terrorist weapon of choice," However, "while [Osama] bin Laden may have his finger on the trigger, his grandson may have his finger on the mouse."[476] Tomorrow's terrorist may be able to do more damage with a keyboard than with a bomb.[477]

Developing effective law enforcement or national security policies to deal with cyber threats is a national priority. However, the private sector must undertake most of the responsibility for fixing weaknesses in key Internet assets. We must understand that it is impossible to eliminate all vulnerabilities. Finally, terrorism is not only a criminal activity—it is a military assault on the entire population, and it must be dealt with accordingly.

Fighting Cybercrime, Cyberterrorism, and Cyberwarfare

Within this chapter, I have discussed some basic tenets of information security, security management, cyberterrorism, cyber-attacks, network security, and user security. This is only a brief overview of some basic principles of information security, and I encourage you to develop a

---

[476] Verton, D. (2002, January 7). Critical infrastructure systems face threat of cyberattacks. *Computerworld, 36(2),* p. 8. Retrieved May 15, 2008, from *http://www.computerworld.com/printthis/2002/0,4814,67135,00.html.*

[477] Weimann, G. (2004, December). Cyberterrorism: How real is the threat? United States Institute of Peace Special Report No. 119. Retrieved April 4, 2008, from *http://www.usip.org/pubs/specialreports/sr119.html.*

greater understanding of the principles of information security and how they relate to the security of the organization by reviewing additional, comprehensive information security resources for more thorough coverage of the subject.

Integrated Efforts in Fighting Cyberterrorism

Over the past 20 years, good work has gone on in cyber security and in mitigating vulnerabilities in critical infrastructures. It took some catastrophic events to gain the attention of lawmakers and policy setters. The pendulum currently swings in the right direction for reliability and necessary security to be designed into future systems. Care must be taken to watch opportunists who take unfair advantage of vulnerabilities to support their hidden agendas. Support for strong systems should come from top-down. All governing and non-governing bodies, both private and public, must focus on the needs of protecting CI systems. The bottom line is that we must learn from the mistakes of others.

Large-scope collaborations are relatively new in American history. It is difficult for national, state, and local organizations to work harmoniously together. The Internet presents opportunities for perpetrators to take advantage of others. Initially, the Internet wasn't designed with security in mind. A certain amount of "trust" resides with computer users. Integration of converged systems with CI goes unabated with the guiding goal of efficiency. Careful, thoughtful, and substantive planning must be the priority as we move forward. All private and public (national, state, and local) entities must be part of a collaborative effort to ensure that we move in a direction that maintains our existence.

Contemporary technology advances at an astounding rate. This advancement offers many new ways to implement public infrastructure services. One needs only to look back a couple of decades to see remarkable differences between old and new CI systems. Unfortunately, inherent vulnerabilities in these systems provide opportunity for perpetrators to take advantage of victims. For the Internet to remain a global conduit, the United States must maintain a more global involvement in the policy equation. This requires international and national communities to embrace closely all aspects of cyber security.

Computer Security, Forensics, and Cyberterrorism

Cybercrime, cyberterrorism, and digital forensics are all relatively new phenomena. As such, local and international governments struggle to remain effective in proposing and passing digital crime legislation. Current consensus fosters the spirit of strict processes in the collection and analysis of digital evidence. At local levels, nations strive to bolster legal prosecution of many new types of digital crime. At the international level, nations are realizing the need to cooperate with other nations to mitigate transnational violations. As technology continues to advance, all peoples of the world will be at risk and will need to be involved in mitigating cybercrimes.

Cybercrime, cyberterrorism, and CI protection present many challenges. Over the past 25 years, much work has been performed in these areas. However, moving forward, more work needs to be done. We must establish and understand common definitions at all levels. We must identify and prioritize (by importance) real vulnerabilities. We must establish clear groups of authority, and these groups must recognize and perform their responsibilities. We must establish strong talent pools in the public sector venue. We must provide these employees with genuine support. We must strengthen program and project management with qualified individuals who can cope with diversity. International collaboration must be part of the culture. Finally, a strong effort must be made to handle all legal issues related to cybercrime, cyberterrorism, and CI protection.

Protecting Our Critical Infrastructure

In the past, infrastructures were far less sophisticated than they are today. With passage of time, populations have grown throughout the world. Concurrent with population growth, a rising need exists to provide additional, required infrastructures to support society. Meanwhile, computers grow more powerful and are continually integrated with infrastructures for improvement. In the United States, both private and public sector organizations heavily rely on computers, information systems, and Internet connectivity as primary components of critical information systems. As these infrastructures grow in complexity, they are scrutinized by those who desire to undermine them. As a result, organizations must work hard to remain ahead of maliciousness. Hence, as infrastructures become more complex, so do efforts to guard and keep them operating safely.

The United States and the world have seen dramatic changes in the way people operate and maintain CIs. Many CIs share a close relationship with using computers, systems, and connectivity. If we were to eliminate these new mechanisms, supporting national security and public needs would become difficult and maybe impossible. Critical infrastructure threats loom greater and more numerous than in the past. Going forward with CI protection requires significant investment by public and private organizations. The government must allocate resources that support more robust CIs. They must strive to create an environment that provides the private sector with investment incentives.

Cyberterrorism is a new phenomenon. In the past 30 years, significant changes have transpired in computer and communications technologies. A global connectivity network provides individuals, organizations, municipalities, states, countries, and national bodies with opportunities to create more efficient and effective infrastructures. This evolution is not without it danger, however. Cyberterrorism presents a clear and present danger to users of CIs. Authoritative governing bodies must catch up to these developments using financial investments and appropriate incentives to public and private sector organizations. When we understand how vulnerable and fragile infrastructures can be, we then realize that there is much more work to be done.

Undoubtedly, the United States and the world must recognize this new battleground, which is far different than previous war zones. Critical infrastructures, information systems, telecommunications, computers, and all operations and controls have converged. It's a modern marvel how they integrate and perform in such a transparent and seamless fashion. Mankind relies on critical systems to perform flawlessly. People who inflict suffering on the masses by damaging/destroying CI exist in the world. The United States and other nations must understand CI vulnerabilities. However, a clear understanding is just the start. As we move forward, focus must be given to protecting and sustaining CIs and the reliable services they provide.

Security of Handheld Devices

Mobile devices and communications networks allow access to information at any time and from anywhere. Organizations are rapidly adopting mobile device technology and seeing major productivity benefits of keeping workers fully functional when mobile. However, cybercriminals target the business user as a passageway into corporate data centers.[478] Thus, security for mobile devices and the software that drives them become critical in protecting organizational information.[479]

I hosted a meeting in June 2009 where the topic of the meeting was a discussion about mobile device security. The people in the meeting were CEOs and chief technology officers (CTOs) of organizations both outside and within the United States. All those present were very knowledgeable about mobile device applications and mobile device security. Their companies use mobile devices, and companies they support or sell to also use mobile devices. The organizations they do business with range from very small to very large enterprises including government organizations. Everyone agreed that most organizations, unless they must be compliant due to regulatory requirements, implement mobile security measures as a *reactionary* rather than a *proactive* function. Additionally, many agreed that something major and newsworthy must *first* happen before organizations will truly understand the consequences and take things seriously. For example, one person in the meeting felt that mobile device security will not be taken seriously until someone hacks President Obama's BlackBerry and leaks confidential information.

Protecting an organization's information remains crucial in today's digital information age. In their design and architecture, handheld devices represent complex computers as do the wireless networks in which they operate.[480] Using handheld devices without correct security

---

[478] Hickey, A. R. (2007). *Mobile security breaches inevitable, study says.* Retrieved October 5, 2007, from *http://searchmobilecomputing.techtarget. com/originalContent/0,289142,sid40_gci1272948,00.html.*

[479] Tauschek, M. (2008). *Mobility policies: Security.* Retrieved May 5, 2008, from *http://viewer.bitpipe.com/viewer/viewDocument.do?accessId= 7500007.*

[480] Shih, D., Lin, B., Chiang, H., & Shih, M. (2008). Security aspects of mobile phone virus: A critical survey. *Industrial Management & Data Systems, 108*(4), 478-474.

prevention measures exposes an organization to risks that could potentially cause an expensive security breach or cyber attack. By implementing these guidelines, organizations can take the necessary steps to protect handheld mobile devices, users, and proprietary information.

Conclusion

In conclusion, *Cybercrime, Cyberterrorism, and Cyberwarfare* provides students and researchers with thoughts and ideas for protecting the United States from cybercriminals, cyberterrorists, and cyberwarriors. We must educate ourselves and know our enemies. We must develop, understand, and be prepared to use all of the countersecurity strategies, tactics, plans, and weapons against all criminals, terrorists, warriors bent on destroying the USA. We will be fighting a good cause, which is to protect our families, religions, homes, property, and American way of life. As long as we are fighting for what is right and good, we will never go wrong and be defeated. We will always come out on top.

Chapter 15
Epilogue

Robert T. Uda, MBA, MS

We might as well get used to it. Criminals, terrorists, and hostile warriors will be around for a long time. Terrorists have become the worst they could ever be with wanton killings of innocent civilians . . . men, women, children, the infirm, young, and old. They do not discriminate against gender, politics, religion, rich, poor, educated, ignorant, sexual orientation, and race. They will kill anyone with bullets, bombs, improvised explosive devices (IEDs), beheadings, incineration, torture, rape, dismemberment, and filth. Now, they even use cyber threats.

They will kill anyone who does not believe as they do. They will kill all heathens, unbelievers, hypocrites, infidels, perverts, Americans, Westerners, Christians, and Jews. All they desire is world domination, a one-religion world, and the demise of what they call The Great Satan (or the United States of America). Their desire is to establish a caliphate with Osama bin Laden as the chief Caliphate and leader of the world. Terrorists do not even care about dying. They believe that their god, Allah, will reward them with 72 virgins if they died for their cause. How pathetic can some people become?

Pure and simple . . . terrorists are criminals. They are motivated by their twisted view of Utopia. They feel that their God gives them the right to kill anyone and everyone. Their twisted philosophy, ideology, and religion guide their every action. They set themselves up as judge, jury, and executioner of all those who oppose them. They do not believe in civil rights. They do not believe in human rights and human dignity. They despise the Constitution of the United States and the American Bill of Rights. All they understand is power, threat, fear, intimidation, death, mutilation, coercion, and dominance. They care for nobody except themselves. What a selfish bunch of people . . . if you can call them people. They are worse than animals. They are savages and throwbacks from the Dark Ages. They are evil as evil can be. They are motivated by the Devil.

On the other hand, criminals want to live. They want to enjoy the spoils of their crimes. Cybercriminals are no different. They do not want to destroy everything so they won't receive the benefits of their illegal efforts. Likewise, hostile nations want to conquer the United States and put us in bondage. They want to receive the spoils of war to enjoy. Therefore, criminals and hostile nations and their warriors are not as vicious and evil as are the terrorists of the world. However, we must prepare for and protect our country against criminals and hostile warriors just as we do against terrorists.

The new battlefield of cyberspace presents new threats and challenges for us. All of these threats and challenges will be in addition to the other threats challenges these perpetrators have fought us with in the past. We just need to be prepared for them, for when we are prepared, we shall not fear.

Appendix A
Abbreviations and Acronyms

ABD	All But Dissertation
ACLU	American Civil Liberties Union
AF	Air Force
AFCYBER	Air Force Cyber Command
AFIT	Air Force Institute of Technology
ANSI	American National Standards Institute
ASEAN	Association of Southeast Asian Nations
ASTM	American Society for Testing and Materials
ATC	Air Traffic Control
BBC	British Broadcasting Corporation
BCP	Business Continuity Planning
BECCA	Business Espionage Controls & Countermeasures Association
BOR	Board of Regents botnet robot network
BP	British Petroleum
BU&A	Bob Uda and Associates
C&C	Command and Control
C3I	Command, Control, Computing, & Intelligence
CAGS	Certificate of Advanced Graduate Studies
CCIPS	Computer Crime and Intellectual Property Section
CCRC	Computer Crime Research Center
CCWMD	Citizens' Crime Watch of Miami-Dade
CD	Compact Disk
CEO	Chief Executive Officer
CERT	Computer Emergency Response Team
CERT	Community Emergency Response Team
CHSP	Certified Homeland Security Professional
CI	Critical Infrastructure
CI2RCO	Critical Information Infrastructure Research Coordination Office

CIA	Central Intelligence Agency
CICTE	Inter-American Committee Against Terrorism
CIO	Chief Information Officer
CIP	Critical Infrastructure Protection
CIPP	Critical Infrastructure Protection Program
CM	Certified Manager
CNI	Critical Network Infrastructure
COB	Chairman of the Board
CONUS	Continental United States
CRS	Congressional Research Service
CSCE	Certificate of Successful Completion of Examination
CSCIC	Cyber Security & Critical Infrastructure Coordination
CSI	Container Security Initiative
CSIS	Center for Strategic and International Studies
CSO	hief Security Officer
CY	Calendar Year
DC	District of Columbia
DCSINT	Deputy Chief of Staff Intelligence
DDoS	Distributed Denial of Service
DHS	Department of Homeland Security
DNI	Director of National Intelligence
DNS	Domain Name Service
DNSSEC	Domain Name System Security Protocol
DoD	Department of Defense
DOJ	Department of Justice
DON	Department of the Navy
DoS	Denial of Service
DOS	Denial of Service
DSW	Disaster Service Worker
ECPA	Electronic Communications Privacy Act
Eds.	Editors
EU	European Union
FBI	Federal Bureau of Investigation
FDA	Food and Drug Administration
FPIF	Foreign Policy in focus
FTO	Foreign Terrorist Organization
GAO	General Accountability Office

GE	General Electric
Gen.	General
GIG	Global Information Grid
GUI	Graphical User Interface
GWOT	Global War on Terrorism, Global War on Terror
HAZMAT	Hazardous Material
HKP	Hong Kong Police
HSD	Homeland Security Department
HTTPS	Hyper Text Transfer Protocol over SSL
IACIS	International Association of Computer Investigation Specialists
IACSP	International Association for Counterterrorism & Security Professionals
IALEIA	International Association of Law Enforcement Intelligence Analysts
ICPM	Institute of Certified Professional Managers
ICTOA	International Counter Terrorism Officers Association
ID	Identification, Identity
IDS	Intrusion Detection System
IED	Improvised Explosive Device
IEEE	Institute of Electrical & Electronics Engineers
INFOSEC	Information Security
IP	Internet Protocol
IRC	Interhemispheric Resource Center
IRC	Internet Relay Chat
IRS	Internal Revenue Service
ISO	International Organization for Standards
ISO	Information Security Officer
ISP	Internet Service Provider
ISS	Information Systems Security
ISTS	Institute for Security Technology Studies
IT	Information Technology
ITU	International Telecommunication Union
IW	Information Warfare
KA	Key Asset
KCC	Korean Communications Commission, South
KCI	Kentucky Critical Infrastructure
LLC	Limited Liability Company

Lt. Gen.	Lieutenant General
MA	Master of Arts
MBA	Master of Business Administration
MIPT	Memorial Institute for the Prevention of Terrorism
MIS	Management Information System
MLA	Mutual Legal Assistance
MS	Master of Science n.d. no date
NASCIO	National Association of State Chief Information Officers
NCSD	National Cyber Security Division
NCU	Northcentral University
NDIA	National Defense Industrial Association
NGO	Non-Governmental Organization
NIHS	National Institute for Hometown Security
NIPC	National Infrastructure Protection Center
NIPP	National Infrastructure Protection Plan
NIST	National Institute of Standards & Technology
NJ	New Jersey
NJIT	New Jersey Institute of Technology
NLC	National League of Cities
NLR	National Lambda Rail
NMSCO	National Military Strategy for Cyberspace Operations
NRC	National Research Council
NSA	National Security Agency
NYS	New York State
OAS	Organization of American States
OMB	ffice of Management and Budget
OS	Operating System
PAN	Personal Area Network
PC	Personal Computer
PCCIP	President's Commission on Critical Infrastructure Protection
PCCW	Pacific Century Cyberworks
PDA	Personal Digital Assistant
PDD	Presidential Decision Directive
PhD	Doctor of Philosophy
PI	Principal Investigator
PIN	Personal Identification Number

PKI	Public Key Infrastructure
PLA	People's Liberation Army
POC	Point of Contact
QOL	Quality of Life
R&D	Research and Development
RAM	Random Access Memory
Ret.	Retired
S&T	Science and Technology
SAE	Society of Automotive Engineers
SAFER	Strategic Actions for Emergency Response
SANS	SysAdmin, Audit, Network, Security
SCADA	Supervisory Control and Data Acquisition
SEI	Software Engineering Institute
SIG	Statica Intelligence Group
SMS	Short Message Service
SSA	Sector-Specific Agency
SSL	Secure Socket Layer
SSP	Sector-Specific Plan
STAR	School Terrorism Awareness & Response
SysAdmin	System Administrator
telcos	telephone companies
U.S.	United States
UCLA	University of California at Los Angeles
UK	United Kingdom
UN	United Nations
USA	United States of America
USAF	United States Air Force
US-CERT	United States Computer Emergency Response Team
USCG	United States Coast Guard
USDCESO	Unified San Diego County Emergency Services Organization
USN	United States Navy
WBIED	Waterborne Improvised Explosive Device
WiFi	Wireless Fidelity

Appendix B
Glossary of Terms

Term	Definition	Source
Adware	Any software application in whichx advertising banners are displayed while the program is running. The authors of these applications include additional code that delivers the ads, which can be viewed through pop-up windows or through a bar that appears on a computer screen. The justification for adware is that it helps recover programming development cost and helps to hold down the cost for the user. *Note*—Adware has been criticized because it usually includes code that tracks a user's personal information and passes it on to third parties without the user's authorization or knowledge. This practice is called **spyware**.	DCSINT
Bot	A malicious program that, upon being installed onto a computer system, allows the attacker to enslave the system into a network of similarly affected systems known as a botnet. The individual computers in a botnet may also be referred to as a bot or a zombie. A special type of bot known as an IRCBot is a program that	F-Secure

Term	Definition	Source
	connects to an Internet Relay Chat (IRC) channel as a normal user or a botnet. The term "bot" is also used in more general situations for programs that perform automated tasks such as scanning web pages, calculating statistics, and so on. Such programs are generally not considered malicious.	
Botnet	A botnet (a portmanteau formed from the words robot and network) is a network of bot-infected computers that can be remotely controlled from a command-and-control (C&C) server. Each infected computer is known as a zombie computer or zombie. An attacker or group of attackers harness the collective resources of a botnet to perform major malicious actions such as sending millions of spam e-mails, performing a DDoS attack, and much more.	F-Secure
Computer Security	Technological and managerial procedures applied to MIS to ensure the availability, integrity, and confidentiality of information managed by the MIS. See also Information Systems Security.	NYS
Cyber-terrorism (FBI)	A criminal act perpetrated by the use of computers and telecommunications capabilities resulting in violence, destruction, and/or DoS to create fear by causing confusion and uncertainty within a given population with the goal of influencing a government or population to conform to a particular political, social, or ideological agenda.	DCSINT

Term	Definition	Source
Denial of Service (DoS)	An attack designed to disrupt network service, typically by overwhelming the system with millions of requests every second, thereby causing the network to slow down or crash	DCSINT
Denial of Service (DoS)	A type of attack conducted over the Internet in which a massive amount of data is sent to a targeted computer system or resource (e.g., a program, website, or network), with the aim of overwhelming and crashing it. DoS attack is typically conducted by a single or small group of computer systems and can be performed in a variety of ways. Even if a DOS attack does not result in the target totally crashing, so much resources may have been diverted to deal with the attack that performance is significantly degraded or other users are unable to use the system or resource until the attack has ended.	F-Secure
Denial of Service (DoS)	The prevention of authorized access to resources or the delaying of time-critical operations. Refers to the inability of a MIS system or any essential part to perform its designated mission either by loss of or degradation of operational capability.	NYS
Distributed Denial of Service (DDoS)	Similar to a DoS attack but involves the use of numerous computers to flood the target simultaneously.	DCSINT
Distributed Denial of Service (DDoS)	A type of attack conducted over the Internet using the combined resources of many computers to bombard and frequently	F-Secure

Term	Definition	Source
	crash a targeted computer system or resource (e.g., a program, website, or network). There are various types of DDoS attacks, which vary based on how the attack is conducted. DDoS attacks are often performed by botnets as the combined resources of all the computers in the network can generate a terrific amount of data, enough to overwhelm most targets' defenses within seconds. Here is an example of how a DDoS attack is conducted: an attacker exploits an vulnerability in a computer system and makes it the DDoS master using Remote Control Software. Later, the intruder uses the master system to identify and manage zombies that perform the attack.	
Encryption	A method of scrambling or encoding data to prevent unauthorized users from reading or tampering with the data. Only individuals with access to a password or key can decrypt and use the data. The data can include messages, files, folders, or disks.	Symantec
Firewall	A barrier to keep destructive forces away from your property.	DCSINT
Firmware	Equipment or devices within which computer programming instructions necessary to the performance of the device's discrete functions are electrically embedded in such a manner that they cannot be electrically altered during normal device operations.	NYS

Term	Definition	Source
Hack	Any software in which a significant portion of the code was originally another program. Many hacked programs simply have the copyright notice removed. Some hacks are done by programmers using code they have previously written that serves as a boilerplate for a set of operations needed in the program they are currently working on. In other cases it simply means a draft. Commonly misused to imply theft of software. See also Hacker.	NYS
Hacker	Advanced computer users who spend a lot of time on or with computers and work hard to find vulnerabilities in IT systems.	DCSINT
Hacker	A person who can break into other persons' computers through the Internet and steal files and other private data to use them for personal gain or illegal purposes.	F-Secure
Hacker	Common nickname for an unauthorized person who breaks into or attempts to break into a MIS by circumventing software security safeguards. Also, commonly called a "cracker." See also Intruder and Hack.	NYS
Hack Tool	Tools that can be used by a hacker or unauthorized user to attack, gain unwelcome access to, or perform identification or fingerprinting of your computer. While some hack tools may also be valid for legitimate purposes, their ability to facilitate unwanted access	Symantec

Term	Definition	Source
	makes them a risk. Hack tools also generally (1) attempt to gain information on or access hosts surreptitiously utilizing methods that circumvent or bypass obvious security mechanisms inherent to the system it is installed on and/or (2) facilitate an attempt at disabling a target computer, thereby preventing its normal use. One example of a hack tool is a keystroke logger—a program that tracks and records individual keystrokes and can send this information back to the hacker. Also applies to programs that facilitate attacks on third-party computers as part of a direct or distributed DoS attempt.	
Hactivist	These are combinations of hackers and activists. They usually have a political motive for their activities and identify that motivation by their actions such as defacing opponents' websites with counter-information or disinformation.	DCSINT
Information Systems Security (ISS)	The protection of information assets from unauthorized access to or modification of information whether in storage, processing, or transit and against the DoS to authorized users or the provision of service to unauthorized users including measures necessary to detect, document, and counter such threats. See also Computer Security.	NYS
Information Security (INFOSEC)	The protection of information systems against unauthorized access to or modification of information, whether in storage, processing, or transit, and	NYS

Term	Definition	Source
	against the DoS to authorized users or the provision of service to unauthorized users including measures necessary to detect, document, and counter such threats. INFOSEC reflects the concept of the totality of MIS security.	
Intruder	An individual who gains or attempts to gain unauthorized access to a computer system or to gain unauthorized privileges on that system. See also Hacker.	NYS
Intrusion Detection	A security service that monitors and analyzes system events to find and provide real-time or near-real-time attempt warnings to access system resources in an unauthorized manner. This is the detection of break-ins or break-in attempts by reviewing logs or other information available on a network.	Symantec
Logic Bomb	A program routine that destroys data by reformatting the hard disk or randomly inserting garbage into data files.	DCSINT
Malicious Code	Software or firmware that is intentionally included in a MIS for an unauthorized purpose. See also Trapdoor, Trojan Horse, Virus, and Worm.	NYS
Malware	Short for **mal**icious sof**tware**. Software designed specifically to damage or disrupt a system such as a virus or a Trojan Horse.	DCSINT

Term	Definition	Source
Malware	Malware is a category of malicious code that includes viruses, worms, and Trojan horses. Destructive malware utilizes popular communication tools to spread, including worms sent through email and instant messages, Trojan horses dropped from websites, and virus-infected files downloaded from peer-to-peer connections. Malware also seeks to exploit existing vulnerabilities on systems making their entry quiet and easy.	Symantec
Management Information System (MIS)	A MIS is an assembly of computer hardware, software, and/or firmware configured to collect, create, communicate, compute, disseminate, process, store, and/or control data or information. Examples include information storage and retrieval systems, mainframe computers, minicomputers, personal computers and workstations, office automation systems, automated message processing systems (AMPSs), and those supercomputers and process control computers (e.g., embedded computer systems) that perform general purpose computing functions.	NYS
Pharming	A type of social engineering attack in which a fraudulent website is used to trick a user into giving out their sensitive personal information such as their banking or e-mail account details. A pharming attack typically depends on "DNS poisoning," which involves seeding the user's hosts file or a DNS server with false information. In this case, the DNS poisoning tactic redirects users	F-Secure

Term	Definition	Source
	from a legitimate website to a copycat website under the attacker's control. Any information the user enters in the malicious website is then compromised. A pharming attack may also be used in conjunction with a 'phishing'attempt. In this case, a misleading message leads the unsuspecting user to the malicious website. Pharming is pronounced the same as "farming."	
Phishing	A form of criminal activity using social engineering techniques. An attempt to fraudulently acquire sensitive information such as passwords, social security numbers, and credit card details by masquerading as a trustworthy person or business in apparent, official electronic communications. Examples are fraudulent emails from know banks, internet sites, etc. asking for information. "Spear Phishing"is phishing used to target a specific group of people or site in an effort to gain a specific piece of information.	DCSINT
Phishing	A type of social engineering attack in which fraudulent communications are used to trick the user into giving out sensitive information such as passwords, account information, and other details. A phishing attack usually involves a fake communication, supposedly from a trusted corporation/institution, that uses an alarming pretext such as "restoring access to a bank account" to pressure the user into providing their sensitive details. The communication is most commonly done via e-mail, but phishing attacks	F-Secure

Term	Definition	Source
	by instant messages and SMSs are also known. Sophisticated attempts direct users to a seemingly-legitimate website, which is actually under the attacker's control. Any information the user enters in the malicious website is then compromised. Phishing is pronounced the same as "fishing." Phishing is a criminal activity in many jurisdictions.	
Phishing	Phishing is essentially an online con game and phishers are nothing more than tech-savvy con artists and identity thieves. They use SPAM, malicious websites, email messages, and instant messages to trick people into divulging sensitive information, such as bank and credit card accounts.	Symantec
Phreaks	A term used to describe telephone hackers.	DCSINT
Risk	A threat that exploits a vulnerability that may cause harm to one or more assets.	Symantec
Risk Analysis	The process of identifying security risks, determining their magnitude, and identifying areas needing safeguards. An analysis of an organization's information resources, its existing controls, and its remaining organizational and MIS vulnerabilities. It combines the loss potential for each resource or combination of resources with an estimated rate of occurrence to establish a potential level of damage in dollars or other assets. See also Risk Assessment and Risk Management.	NYS

Term	Definition	Source
	Note: Risk analysis is a part of risk management, which is used to minimize risk by specifying security measures commensurate with the relative values of the resources to be protected, the vulnerabilities of those resources, and the identified threats against them. The method should be applied iteratively during the system life-cycle. When applied during the implementation phase or to an operational system, it can verify the effectiveness of existing safeguards and identify areas in which additional measures are needed to achieve the desired level of security. Numerous risk analysis methodologies and automated tools exist to support them.	
Risk Assessment	Process of analyzing threats to and vulnerabilities of an MIS to determine the risks (potential for losses) and using the analysis as a basis for identifying appropriate and cost-effective measures. See also Risk Analysis and Risk Management.	NYS
Risk Assessment	The computation of risk. Risk is a threat that exploits some vulnerability that could cause harm to an asset. The risk algorithm computes the risk as a function of the assets, threats, and vulnerabilities. One instance of a risk within a system is represented by the formula (Asset * Threat * Vulnerability). Total risk for a network equates to the sum of all the risk instances.	Symantec
Risk Management	The total process of identifying, measuring, controlling, and eliminating or minimizing	NYS

Term	Definition	Source
	uncertain events that may affect system resources. Risk management encompasses the entire system life-cycles and has a direct impact on system certification. It may include risk analysis, cost/benefit analysis, safeguard selection, security test & evaluation, safeguard implementation, and system review. See also Risk Analysis and Risk Assessment.	
Sniffer	A program designed to assist hackers/and or administrators in obtaining information from other computers or monitoring a network. The program looks for certain information and can either store it for later retrieval or pass it to the user.	DCSINT
Spam	The unsolicited advertisements for products and services over the Internet, which experts estimate to comprise roughly 50 percent of all sent e-mail.	DCSINT
Spam	Flooding the Internet with many copies of the same message in an attempt to force the message on people who would not otherwise choose to receive it.	F-Secure
Spoofing, E-mail	A method of sending e-mail to a user that appears to have originated from one source when it actually was sent from another source.	DCSINT
Spyware	Any technology that gathers information about a person or organization without their knowledge. Spyware can get into a computer as a software virus or as the result of installing a new program. Software	DCSINT

Term	Definition	Source
	designed for advertising purposes, known as adware, can usually be thought of as spyware as well because it invariably includes components for tracking and reporting user information.	
Spyware	A software program that aids in gathering information about a person or organization without their knowledge and can relay this information back to an unauthorized third party.	F-Secure
Spyware	Spyware is any software package that tracks and sends personally identifiable information or confidential information to third parties. Personally identifiable information is information that can be traced to a specific person such as a full name. Confidential information includes data that most people would not be willing to share with someone and includes bank details, credit card numbers, and passwords. Third parties may be remote systems or parties with local access.	Symantec
Threat	An adversary having the intent, capability, and opportunity to cause loss or damage—(from DoD Directive 3020.40, 19 August 2005).	DCSINT
Threat	An event, process, activity (act), substance, or quality of being perpetuated by one or more threat agents, which, when realized, has an adverse effect on organization assets resulting in losses attributed to (1) direct loss, (2) related direct loss, (3) delays or denials, (4) disclosure of sensitive	NYS

Term	Definition	Source
	information, (5) modification of programs or data bases, and (6) intangible, i.e., good will, reputation, etc.	
Threat	A circumstance, event, or person with the potential to cause harm to a system in the form of destruction, disclosure, data modification, and/or Denial of Service (DoS).	Symantec
Threat Agent	Any person or thing, which acts or has the power to act to cause, carry, transmit, or support a threat. See also Threat.	NYS
Threat Assessment	The severity rating of the virus, worm, or Trojan horse. The threat assessment includes the damage that this threat causes, how quickly it can spread to other computers (distribution), and how widespread the infections are known to be (wild).	Symantec
Trapdoor	A secret undocumented entry point into a computer program used to grant access without normal methods of access authentication. See also Malicious Code.	NYS
Trojan Horse	A program or utility that falsely appears to be a useful program or utility such as a screen saver. However, once installed, it performs a function in the background such as allowing other users to have access to your computer or sending information from your computer to other computers.	DCSINT

Term	Definition	Source
Trojan Horse	A software program with hidden destructive functionality that can damage data, hijack computers, sniff files or network, or steal confidential information.	F-Secure
Trojan Horse	A computer program with an apparently or actually useful function that contains additional (hidden) functions that surreptitiously exploit the legitimate authorizations of the invoking process to the detriment of security. See also Malicious Code and Threat Agent.	NYS
Virus	A software program, script, or macro that has been designed to infect, destroy, modify, or cause other problems with a computer or software program.	DCSINT
Virus	A software program that replicates by attaching itself to another object.	F-Secure
Virus	Code imbedded within a program that causes a copy of itself to be inserted in one or more other programs. In addition to propagation, the virus usually performs some unwanted function. Note that a program need not perform malicious actions to be a virus. It needs only to infect other programs. See also Malicious Code.	NYS
Vulnerability	A flaw or security loophole that may allow other users, applications, or attackers to affect a program or system without the user's authorization or knowledge. A vulnerability can be a flaw in a program's fundamental design, a bug in its code that allows improper usage of the program, or	F-Secure

Term	Definition	Source
	simply weak security practices that allow attackers to access the program without directly affecting its code. Patching a vulnerability requires the program vendor to create a patch—or code to rectify the flaw or loophole—and distribute it to all users in order to protect the system from exploitation.	
Vulnerability	A (universal) vulnerability is a state in a computing system (or set of systems) which allows an attacker to do any or combination of the following: (1) execute commands as another user, (2) access data that is contrary to the specified access restrictions for that data, (3) pose as another entity, and/or (4) conduct a DoS.	Symantec
Vulnerability Assessment	The identification and quantification of a system's technical and environmental vulnerabilities.	Symantec
Vulnerability Management	The practice of identifying and removing weaknesses that can be used to compromise the confidentiality, integrity, or availability of a computer information asset. Vulnerability management is a preventative information security practice that identifies and removes weaknesses before they can be used to compromise a computer information asset.	Symantec
Worm	A destructive software program containing code capable of gaining access to computers or networks and, once within the computer or network, causing that computer or network harm by deleting,	DCSINT

Term	Definition	Source
	modifying, distributing, or otherwise manipulating the data.	
Worm	A software program that replicates independently by sending itself to other systems.	F-Secure
Worm	A computer program that can replicate itself and send copies from computer to computer across network connections. Upon arrival, the worm may be activated to replicate and propagate again. In addition to propagation, the worm usually performs some unwanted function. See also Malicious Code.	NYS
Zombie	A computer or server that has been basically hijacked using some form of malicious software to help a hacker perform a DDoS attack.	DCSINT
Zombie	A computer system or server that is connected to the Internet and has been infected with specialized malware that allows the attacker to use the machine's resources. A bot-infected system will often be harnessed into a botnet, or a collection of similarly infected machines. The collective resources of the botnet can be used to perform a variety of malicious actions including launching DDoS attacks or sending out spam.	F-Secure

References

DCSINT (2006, August 10). Critical infrastructure threats and terrorism. *DCSINT Handbook No. 1.02.* Fort Leavenworth, KS: US Army Training and Doctrine Command, Deputy Chief of Staff for Intelligence, Assistance Deputy Chief of Staff for Intelligence—Threats. Retrieved July 11, 2009, from *http://www.fas.org/irp/threat/terrorism/sup2.pdf.*

F-Secure Corp. (n.d.). *F-Secure ABC malware glossary and Terminology.* F-Secure Website. Retrieved July 11, 2009, from*http://www.f-secureusa. com/abc/virus_information/* and *http://www.f-secure.com/en_US/.*

NYS (n.d.). Automated information systems security policy glossary. New York State (NYS) Office of Cyber Security & Critical Infrastructure Coordination (CSCIC). NYS uses the glossary from the U.S. Customs Service. Retrieved July 11, 2009, from *http://www.cscic.state.ny.us/lib/ glossary/.*

Symantec Corp. (n.d.). *Glossary.* Symantec Security Response Website. Retrieved October 23, 2008, from *http://www.symantec.com/business/ security_response/glossary.jsp.*

About the Author

Robert T. (Bob) Uda, Certified Manager (CM), currently serves as vice president of the West Coast Division of SIG Homeland Security, LLC, headquartered in Florham Park, New Jersey. Bob furthermore serves as president and principal consultant of Bob Uda and Associates (BU&A), a counterterrorism research firm. Bob previously served as chairman, president, and CEO of Apollo Systems Technology, Inc., a system security engineering company in Canyon Country, California. He served in the USAF for over eight years and worked in the aerospace/defense industries for over a quarter century.

Bob earned BS degrees in aerospace engineering from the University of Oklahoma and in general business from Regents College (now called Excelsior College) of Albany, New York. He further earned an MS degree in astronautics from the Air Force Institute of Technology (AFIT) and an MBA from the University of La Verne (California). Additionally, he received a diploma in The Executive Program in Management from the UCLA Graduate School of Management. Bob currently pursues a PhD in Business Administration with specialization in Homeland Security from Northcentral University (NCU) located in Prescott Valley, Arizona. He is doing his dissertation research on exploring the factors contributing to potential improvised explosive device (IED) attacks on the U.S. homeland.

Bob has prepared over 40 publications plus 16 books. He has written/published four counterterrorism books including (1) *Principles of Asymmetrical Warfare: How to Beat Islamo-fascists at Their Own Game*, (2) *Terrorism and Counterterrorism: Victory over Islamo-fascist Jihadists*, (3) *Combating Terrorists in the USA: Protecting the CONUS from Terrorists*, and (4) *Cybercrime, Cyberterrorism, and Cyberwarfare: Crime, Terror, and War without Weapons*. Currently, he is preparing a book titled *Those Blasted IEDs! Defeating the Deadly Improvised Explosive Device*.

Bob is certified in the Community Emergency Response Team (CERT) Program, is a member of the City of San Marcos (CA) CERT, and is a disaster service worker (DSW) with the Unified San Diego County Emergency Services Organization. He earned a Certificate of Successful Completion of Examination (CSCE) Amateur Radio "Technician" License with call sign KI6SQW. Bob also holds the Certified Homeland Security Professional (CHSP) certification with specialization in Counter-Terrorism from SIG Homeland Security, LLC. Additionally, he received the Certificate of Advanced Graduate Studies (CAGS) in Homeland Security from NCU.

Bob holds memberships in the International Association for Counterterrorism & Security Professionals (IACSP), International Association of Law Enforcement Intelligence Analysts, Inc. (IALEIA), International Counter Terrorism Officers Association (ICTOA), InfraGard San Diego Members Alliance, Business Espionage Controls & Countermeasures Association (BECCA), and National Defense Industrial Association (NDIA). Bob is also a member of Alpha Phi Sigma, the honorary society for Criminal Justice and Homeland Security, and was awarded the 2009 Provost Scholarship Award ($500) and Certificate of Achievement from NCU.

Bob received the Honorable Mention Award in the Maritime Security Expo and NCU paper competition. His paper is titled "Detecting and Defeating Waterborne Improvised Explosive Devices (WBIEDs) Onboard Small Vessels." He also published an article on "Thoughts of a Volunteer Hostage" in vol. 7, issue 3, of the *Counter-Terrorism Quarterly*, the official newsletter of the ICTOA. Further, he serves as chairman of the board (COB) of the Institute of Certified Professional Managers (ICPM) Board of Regents (BOR) headquartered at the James Madison University in Harrisonburg, Virginia.

Bob participated in the following natural and manmade disaster training: CERT Academy, American Red Cross Shelter Operations Training, North San Diego Countywide CERT Training Day Exercise, San Onofré Nuclear Generating Station (SONGS) Reception & Decontamination Exercise and served as an evacuee, San Diego InfraGard Chapter Private Sector Counterterrorism Awareness Workshop, HALO Corporation Bicoastal Counter-terrorism Seminar, CERT Training as part of the Golden Guardian Exercise, San Diego InfraGard Chapter Security Officer's Workshop, HALO Corporation Bicoastal Counterterrorism Seminar with Dr. Michael Scheuer, and the InfraGard Los Angeles School Terrorism Awareness & Response (STAR) Exercise 2009 and served as a volunteer hostage.

Bob and his wife, the former Karen Elizabeth Rowland of Circleville, Ohio, sired two sons, a daughter, and currently are proud grandparents of six grandchildren. They live in San Marcos, California. You can contact Bob Uda by e-mail at *robert.uda@sighls.org*.

Contributing Authors

Rhonda Chicone is the CTO and vice president of Engineering for Notify Technology Corporation, a Silicon Valley (California) based software company that specializes in secure wireless mobile software products and services. Rhonda pioneered Notify's R&D Center located in Canfield, Ohio, and is responsible for all aspects of Notify's technology and product development. She also is an adjunct professor at South University Online and Minot State University. She holds an MS degree in Technology from Kent State University and a BS degree in Computer Science from Youngstown State University. Rhonda will soon complete her PhD in Business Administration with specialization in Applied Computer Science. She resides in Canfield, Ohio with her daughter, Deana.

William Shervey began his technical career in the mid-1970s as a radar technician in the United States Coast Guard (USCG). Since then, he has worked with companies such as RCA, GE, Computer Sciences, and Raytheon. He and his wife, Debbe, started and ran their own small business for nine years. Bill holds a BS in system science, MS in information systems, and an MBA. A dedicated, lifelong learner, Bill will soon complete his PhD in Business Administration with specialization in Computer and Information Security. As of this writing, Bill works as an information security analyst. His passion is educating computer users through information security awareness training.

Darrin L. Todd, MS, is an information security engineer with a quarter century of information technology (IT) experience. Taking his first programming class as a teen in 1981, he soon realized that he understood and enjoyed computers. He joined the Air Force in 1985 and spent most of his career working as an IT specialist, supervisor, and superintendent while serving a dozen assignments around the world. He eventually focused on information security. Today, Darrin Todd is a doctoral student and a contractor/consultant working for various government organizations and agencies in the Washington, DC, metro area. These organizations include the Department of Defense and the Department of Homeland Security.

Index

www.ingramcontent.com/pod-product-compliance
Lightning Source LLC
Chambersburg PA
CBHW051226050326
40689CB00007B/823